高等学校专业英语教材

地质英语读写教程

邹 灿 张 远 主编

任怡霖 李 娟 林晨军 副主编

电子工业出版社.

Publishing House of Electronics Industry

北京 · BEIJING

内 容 简 介

本书涵盖地质专业的几大主要基础板块，遵循专业英语课程教学的系统性及完整性，按照其专业基础的编排顺序层层递进，突出了地质类专业的理论性、应用性和前瞻性；以普通地质学的基本概念及基础理论为主，涵盖了地质专业基础、地质构造、地质沉积、矿物岩石等内容。全书共 8 个单元，每个单元提供两篇与单元主题相关的文章并设置了阅读理解、翻译、写作等不同侧重的练习题，一方面培养学生的英语应用能力，有利于学生运用英语技能学习专业基础知识，为进一步的专业英语学习打下坚实的基础；另一方面通过学习地球的演变、发展及矿物的形成，使学生有意识地爱护人类赖以生存的地球家园，努力挖掘造福人类的自然资源。

本书既可作为高等院校地质类相关专业本科生和研究生的专业英语阅读与写作教材，也可作为石油地质工程专业技术人员培训、学习的参考书。

图书在版编目（CIP）数据

地质英语读写教程 / 邹灿，张远主编. —北京：电子工业出版社，2024.2

ISBN 978-7-121-46848-3

Ⅰ. ①地…　Ⅱ. ①邹…　②张…　Ⅲ. ①地质学—英语—教材　Ⅳ. ①P5

中国国家版本馆 CIP 数据核字（2023）第 239516 号

责任编辑：戴晨辰　　特约编辑：张燕虹
印　　刷：河北鑫兆源印刷有限公司
装　　订：河北鑫兆源印刷有限公司
出版发行：电子工业出版社
　　　　　北京市海淀区万寿路 173 信箱　邮编：100036
开　　本：787×1 092　1/16　印张：9.75　字数：324 千字
版　　次：2024 年 2 月第 1 版
印　　次：2024 年 2 月第 1 次印刷
定　　价：49.00 元

凡所购买电子工业出版社图书有缺损问题，请向购买书店调换。若书店售缺，请与本社发行部联系，联系及邮购电话：(010) 88254888，88258888。

质量投诉请发邮件至 zlts@phei.com.cn，盗版侵权举报请发邮件至 dbqq@phei.com.cn。

本书咨询联系方式：dcc@phei.com.cn。

前　言

Preface

    随着我国地质领域对外交流与合作的日益增长，为提高地质专业的学习者、科技工作者、涉外人员专业英语的阅读、翻译和写作能力，我们精心组织编写了《地质英语读写教程》。本书以地质学的基本概念及基础理论为主，涵盖了地质专业基础、地质构造、地质沉积、矿物岩石等内容。全书遵循专业英语课程教学的系统性及完整性，并遵循由浅入深、循序渐进的教学规律，使专业英语教学与专业课教学相匹配，同时突出了地质类专业的理论性、应用性和前瞻性。因本书所有课文均选自国外出版的专业书籍，参考与引用文献主要源自正式出版的英、美地质专业原著，故翔实可靠、语言规范。全书共 8 个单元，每个单元围绕一个主题，包括 Text A 和 Text B 两篇文章。Text A 后附有 New words、Notes to the text、Understand the text、Translation、Writing skill；Text B 后附有 Notes to the text、Questions for review。每个单元针对科技论文的写作进行了循序渐进的讲解，以帮助读者提高写作技能。

    本书构思独特、实用性强，能满足对专业论文写作的实际需要。为利于教学和自学，书后附有 Glossary（词汇总表）和 Keys to Exercises（练习题答案）。

    对于本书的配套课件，读者可登录华信教育资源网（www.hxedu.com.cn）免费下载；对于其他拓展学习资源，读者可联系作者（邮箱：444538927@qq.com）获取。

    主编邹灿教授和张远老师（成都理工大学）参与了本书的编写工作并负责了全书的审校工作。副主编任怡霖副教授、李娟副教授（成都理工大学）、林晨军老师（西华大学）参与了本书的编写工作。在编著过程中，地质领域专家、博士生导师王国芝教授对本书给予了指导，英国友人 Charles Gray 审阅了本书英文内容，在此一并表示诚挚的谢意。

　　本书既可作为高等院校地质类相关专业本科生和研究生的专业英语阅读与写作教材，也可作为石油地质工程专业技术人员培训、学习的参考书。

　　由于水平有限，本书难免存在差错或不足之处，敬请专家、老师、同学多提宝贵意见，以便不断改进和完善本书。

编　者

目 录

Contents

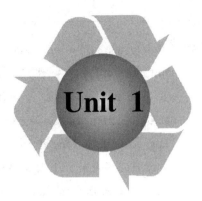

Unit 1

Geology

The Earth

1 The earth is one of the eight planets, which all revolve around[1] the Sun. Of them the earth is neither the largest nor the smallest. The earth is constantly changing. Long, long ago, the earth was not like the earth today. It was only a **molten sphere**. Gradually the molten sphere cooled down. At last the rocks and life appeared on the earth. It's commonly believed that the earth is a huge mass of rock, water, and gases. The general shape of the earth is that of a **flattened** sphere. It is not perfectly round but slightly flattened at the poles. **Artificial** satellites have confirmed that the earth has a **circumference** of approximately 25,000 miles (40,000 km), and it is 7,927 miles (12,756 km) in **diameter** at the **equator** and about 27 miles less, namely about 7,900 miles (12,714 km), in diameter at the poles[2]. This is a small difference amounting to only one-third of one percent when compared to the earth's size[3].

2 The earth is usually divided by scientists into three spheres: the atmosphere, the **hydrosphere** and the **lithosphere**. The solid part of the earth is called the lithosphere.

Actually, we know more about the water and gases covering the earth than we know about the solid earth itself. We can see the surface of the earth, and mines and deep wells tell us a little of what is under the surface. However, the deepest wells drilled by **geologists** from the former Soviet Union extend only about 7 miles below the surface. Thus, compared to the almost 8,000-mile diameter of the earth, this is only a **scratch** on its surface. What is below this 7-mile depth? No one knows for sure, but scientists believe the earth is made up of several layers.

3　The outermost layer, the part we see on the surface of the earth, is the **crust**. This is believed to be about 7 to 30 miles thick in land areas and thinner under the oceans. Geologists have found that the crust is about 4 to 10 miles thick under most of the oceans. The rocks of the crust actually form a thin shell which covers the surface of the globe.

4　The **mantle**, which is about 1,800 miles thick, lies underneath the crust. It is thought to be composed of heavier rocks than the material making up the crust[4]. Because of the great pressure and heat in this layer, the mantle is neither quite a solid nor quite a liquid[5]. In other words, the material making up the mantle is thought to be in a plastic state. The land masses making up the crust are thought to "float" on this plastic layer.

5　The next layer is the outer **core**, nearly 1,400 miles thick. It is believed to be composed mostly of iron and **nickel** in a molten state at a very high temperature. Evidence indicates that the outer core is about twice as dense as the material in the mantle.

6　The innermost layer, the inner core, extends about 800 miles to the earth's center. It also is probably composed of iron and nickel. Because the pressure is nearly 60,000,000 pounds per square inch, the inner core is more like a true solid. The material making up the inner core is thought to be about three or four times as dense as the material making up the mantle[6]. You might think of the earth as being constructed much like a baseball with a two-layered core, a thick layer being around the core (mantle), and a thin skin on the surface (crust)[7].

7　The other two of the three spheres will be introduced in the following. The hydrosphere is the water layer covering the earth. Water and air have been the most important factors in changing the surface of the earth throughout its history. Most of the earth's water is in the oceans, but rivers and lakes are also a part of the hydrosphere. How much water is there on the earth? Scientists have estimated that if all the mountains were leveled off and sea bottom were raised, there would be enough water to cover the earth to a depth of a mile and a half.

8 The atmosphere surrounds the lithosphere and the hydrosphere. The atmosphere is made up of gases, chiefly **nitrogen** and oxygen, surrounding the solid and liquid parts of the earth. In the study of the earth, only a few of these gases will be mainly concerned about. Oxygen is important because it is combined with many other minerals in the earth's crust. Carbon **dioxide**, in water solution, also has helped to form many rocks and minerals. At present, the composition of dry air is well-known as nitrogen (78.08%), oxygen (20.95%), **argon** (0.93%), carbon dioxide (0.03%) and rare gases (0.01%) (Figure 1-1).

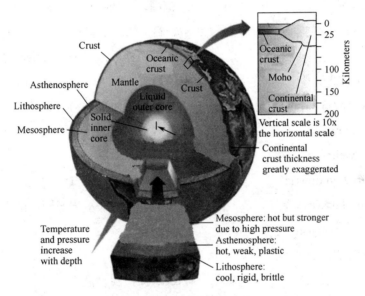

Figure 1-1   The Earth's Interior

9 The water vapor in the earth's atmosphere is responsible for many **geological** changes[8]. It is one of the most essential **agents** by which the earth's surface has been changed over many, many millions of years[9].

10 The earth is a complex, **dynamic** planet that has changed continually since its origin some 4.6 billion years ago. These changes and the present-day features we observe are the result of **interactions** between the various **interrelated** internal and external earth **subsystems** and cycles. In fact, the earth is unique among the planets of our solar system in that it supports life and has oceans of water, a **hospitable** atmosphere, and a variety of

climates[10]. It is ideally suited for life as we know because of a combination of factors, including its distance from the Sun and the evolution of its **interior**, crust, oceans, and atmosphere. Over time, changes in the earth's atmosphere, oceans, and to some extent, its crust have been influenced by life processes. In turn, these physical changes have affected the evolution of life.

New words

molten /ˈməultən/	*adj.* 熔化的
	v. 熔化，溶解，变软
sphere /sfɪə(r)/	*n.* 范围，领域；球，球体
flattened /ˈflætnd/	*adj.* 扁平的；平缓的
	v. 把……弄平；变平（flatten 的过去分词）
artificial /ˌɑːtɪˈfɪʃl/	*adj.* 人造的
circumference /səˈkʌmf(ə)r(ə)ns/	*n.* 圆周，周围
diameter /daɪˈæmɪtə/	*n.* 直径
equator /iˈkweitə/	*n.* 赤道
hydrosphere /ˈhaidrəsfiə/	*n.* 水圈，水界，水气
lithosphere /ˈlɪθəsfiə/	*n.* ［地物］［地质］岩石圈；陆界
geologist /dʒɪˈɔlədʒɪst/	*n.* 地质学家
scratch /skrætʃ/	*n.* 擦痕，划痕
	vt.&vi. 抓，刮，搔
crust /krʌst/	*n.* 地壳；外壳，坚硬的外壳，面包皮
mantle /ˈmæntl/	*n.* 地幔；罩；盖层
core /kɔː/	*n.* ［地］地核；岩心
nickel /ˈnikl/	*n.* ［地］自然镍；镍
nitrogen /ˈnaitrədʒən/	*n.* 氮
dioxide /daɪˈɒksaɪd/	*n.* 二氧化物
argon /ˈɑːgɒn/	*n.* ［化］氩（18 号化学元素，符号为 Ar）
geological /dʒɪəˈlɒdʒɪkəl/	*adj.* 地质（学）的
agent /ˈeɪdʒ(ə)nt/	*n.* 营力，作用力；因素；作用物

dynamic /daɪˈnæmɪk/	*adj.* 动态的；充满活力的，精力充沛的；不断变化的，充满变数的
interaction /ˌɪntərˈækʃən/	*n.* 交互影响，相互作用
interrelated /ˌɪntərɪˈleɪtɪd/	*adj.* 相互关联的
	v. 相互关联（影响）（interrelate 的过去式和过去分词）
subsystem /ˈsʌbsɪstəm/	*n.* 子系统，次系统；亚晶系
hospitable /hɒˈspɪtəbl, ˈhɒspɪtəbl/	*adj.* 好客的；热情友好的；（环境）舒适的
interior /ɪnˈtɪəriə(r)/	*n.* 内部，内陆
	adj. 内部的，本质的；国内的

Notes to the text

(1) revolve around：环绕，围绕

(2) Artificial satellites have confirmed that the earth has a circumference of approximately 25,000 miles (40,000 km), and it is 7,927 miles (12,756 km) in diameter at the equator and about 27 miles less, namely about 7900 miles (12,714 km), in diameter at the poles.

that 引导 confirm 的宾语从句。

译文：人造卫星已经证实，地球的周长约为 25 000 英里（40 000 千米），赤道直径为 7 927 英里（12 756 千米），两极间的直径约为 7 900 英里（12 714 千米），比赤道直径小约 27 英里。

(3) This is a small difference amounting to only one-third of one percent when compared to the earth's size.

amount to：总共达到（总计，等于），相当于；意味着

compared to：与……相比；比作

译文：两者差别很小，只相当于地球大小的 0.3%。

(4) It is thought to be composed of heavier rocks than the material making up the crust.

be composed of：由……构成；making up…是现在分词做 the material 的后置定语。

译文：人们认为构成它的岩石比组成地壳的物质更重。

(5) Because of the great pressure and heat in this layer, the mantle is neither quite a solid nor quite a liquid.

译文：由于该圈层巨大的压力和热量，地幔既非完全的固态也非纯粹的液态。

(6) The material making up the inner core is thought to be about three or four times as dense as the material making up the mantle.

这里的两个 making up 均是现在分词分别做各自前面的 the material 的后置定语。

译文：人们认为，组成内核的物质密度大约是组成地幔的物质密度的 3 倍或 4 倍。

(7) You might think of the earth as being constructed much like a baseball with a two-layered core, a thick layer being around the core (mantle), and a thin skin on the surface (crust).

think...as：把……认为，把……想象成

a thick layer being..., and a thin skin...：这是两个独立主格结构，因 a thin skin...后面部分与前面结构一致，故省略了 being。

译文：你可以将地球的结构想象成一个具有双层内核的棒球，包绕着核的厚层为地幔，表层的薄层为地壳。

(8) The water vapor in the earth's atmosphere is responsible for many geological changes.

be responsible for：担负(对……负责)；……是……的原因

译文：地球大气层里的水蒸气是引发很多地质变化的原因。

(9) It is one of the most essential agents by which the earth's surface has been changed over many, many millions of years.

by which 做 agents 的定语。

译文：它是引起地球表面数百万年来变化的最重要因素之一。

(10) In fact, the earth is unique among the planets of our solar system in that it supports life and has oceans of water, a hospitable atmosphere, and a variety of climates.

in that：因为；既然；在于

oceans of：很多的，无限的

译文：事实上，地球是我们太阳系中独一无二的行星，因为它支撑生命、有无尽的海洋、宜人的大气和多变的气候。

Understand the text

Answer the following questions according to the passage you have read.

(1) What's the general shape of the earth?

(2) How many parts is the earth usually divided into?

(3) What's the lithosphere? And what do you know about it?

(4) How many layers is the earth made up of and what are they?

(5) Is the crust thought to be composed of heavier rock than the material making up the mantle?

(6) Why is the mantle thought to be in a plastic state?

(7) What is the core believed consisting of ?

(8) What is the hydrosphere?

(9) What is one of the most essential agents responsible for many geological changes?

(10) What have the changes and the present-day features we observe today resulted from?

Translation

1. Translate the following sentences into Chinese.

(1) Thus, compared to the almost 8,000-mile diameter of the earth, this is only a scratch on its surface.

_____

_____

_____

_____

(2) In other words, the material making up the mantle is thought to be in a plastic state.

_____

_____

_____

_____

(3) Scientists have estimated that if all the mountains were leveled off and sea bottom were raised, there would be enough water to cover the earth to a depth of a mile and a half.

_____

_____

_____

_____

(4) The atmosphere is made up of gases, chiefly nitrogen and oxygen, surrounding the solid and liquid parts of the earth.

_____

_____

_____

(5) Over time, changes in the earth's atmosphere, oceans, and to some extent, its crust have been influenced by life processes.

_____

_____

_____

_____

2. Translate the following passage into English.

　　地球的形状近似于球体。它的周长约为 25 000 英里（40 000 千米），其两极间的直径约为 7 900 英里（12 714 千米），赤道直径为 7 927 英里（12 756 千米）。地球内部的三大圈层是地核、地幔和地壳。地核的直径大约是 4 300 英里（6 900 千米），其主要成分极可能是铁。地核由看似固体的内核和看似液体的外核组成。地幔厚度接近 1 800 英里（2 900 千米），约占地球体积的 84%。由于地核的体积接近（地球体积的）16%，总的来看，地壳实际上只占地球非常小的部分。

_____

_____

_____

_____

_____

_____

_____

_____

_____

Writing skill

Features of EST at Lexicon Level

EST is an acronym（首字母缩写词）, standing for English for Science and Technology and covering the areas of English written for academic and educational purposes and of English written for occupational purpose.

Diction of EST is different from that of literature or everyday usage. Technical words and semi/sub-technical words are commonly used. The other outstanding feature of EST at lexicon level is abbreviation.

Technical words refer to words that can exactly explain definite concepts in a specific field of science and technology. In Text A, an abundant of technical words are involved, such as *hydrosphere, lithosphere, mantle, nitrogen, argon, etc.*

Semi/sub-technical words refer to terms which have multiple meanings, but are given a precise definition for scientific use. There are some of this type of words in Text A, for example, *crust, solid, liquid, core, well, agent.*

Non-technical words refer to those words which are not technical in a strict sense, but tend to occur particularly frequently in EST. For example:

Words preferred in EST	Words used everyday
determine	Find out
convert	change
maximum	greatest
approximately	about
possess	have
come into operation	start to be used

Writing practice

You may try to pick out more examples of these three kinds of diction respectively from both Text A and Text B.

Text B

Science of the Earth

1 Geology is the study of the earth: its physical aspects and its history. It is the study of a volcano erupting in glowing gas, clouds of ash, and streams of molten rock[(1)]. It is the reconstruction of prehistoric[(2)] plants and animals on a landscape drained by rivers flowing into long gone seas – a scene that disappeared hundreds of millions of years ago. It is a concept of land rising from the sea to form lofty mountains that are later worn away to low rolling plains[(3)] – all during an immensity of time. It is a mental picture of the earth's structure from the outer surface to the center 4,000 miles inward, or of the neat arrangement of invisible particles which gives a crystal[(4)] its outer form. In brief, geology is the study of the composition, structure, and history of the earth.

2 The great mass of detail that constitutes geology is classified under a number of subdivisions which, in turn, depend upon the fundamental sciences: physics, chemistry, and biology. In their study of rocks and minerals geologists apply chemical and physical concepts of atoms, molecules[(5)], and crystals. In determining the date when rocks were formed, geologists may use findings on radioactive isotopes[(6)] from atomic physics. In paleontology[(7)], the science of "ancient life", geologists call upon key concepts of biology and on the details of anatomical[(8)] findings to discover the past history of living things as recorded on the rocks.

3 These secrets of the earth are probed in many ways. Waves from the earthquakes travel through the interior to bring out messages about its structure that are written on seismographs[(9)]. Chemical and X-ray analyses reveal the composition of rocks and minerals. Mapping of sands, travels, and boulders[(10)] shows where glaciers[(11)] once moved over the land. Examination of ancient caves and strand lines[(12)] proves that the oceans have been hundreds of feet deeper than they are today, while submarine canyons and wave-planed volcanic peaks under deep water present the possibility that the oceans have also been

hundreds, even thousands of feet shallower.

4　While geology has, as its central themes, the materials and structure of the earth and the earth's history, it is overlapped by many other earth-oriented sciences. To name only a few, there are meteorology[13], for the study of the atmosphere; physical geography, for the external features of the globe; oceanography[14], for the earth's waters and their uneven depths; soil science, for its vital and special topic. Then there are branches of geophysics[15] and geochemistry[16], testifying to the sheer bulk of known and knowable matters that have required ever-growing specialization of science in our times. It is a synthesis of the natural sciences: astronomy, biology, chemistry, mathematics, and physics. And, above all, it has something for everyone. Who can experience or even hear about an earthquake or volcanic eruption without wondering about its cause? If you find a sea shell or fish solidly encased in the rock of an inland stream bed[17], or of a high mountain, would you wonder why it was there? Have you ever pondered the jumbled varicolored rocks or multitudinous grains of sand of a shoreline, the gold like glitter of yellow mica[18] in a piece of field stone[19], or the smooth symmetry of a quartz crystal[20]? If these or any of thousand and one phenomena all around us have stimulated so much as a fleeting question in your mind, you have peeked through a door into the world of geology. Anyone can walk through such a door and find treasures limited only by the dimensions of his curiosity and enthusiasm.

Notes to the text

(1) molten rock：熔岩

(2) prehistoric /ˌpriːhɪˈstɒrɪk/ *adj.* 史前的

(3) rolling plain：波状/起伏平原

(4) crystal /ˈkrɪstəl/ *n.* 水晶，晶体

(5) molecule /ˈmɒlɪkjuːl/ *n.* 分子，克分子

(6) radioactive isotope：放射性同位素

(7) paleontology /ˌpælɪɒnˈtɒlədʒɪ/ *n.* 古生物学

(8) anatomical /ænəˈtɒmɪkl/ *adj.* 结构上的，解剖（学）的

(9) seismograph /ˈsaɪzmə(ʊ)grɑːf/ *n.* 地震仪

(10) boulder /ˈbəʊldə(r)/ *n.* 巨砾，漂砾

(11) glacier /ˈglæsiə(r)/ *n.* 冰河，冰川

(12) strand line：滨线；海岸线

(13) meteorology /ˌmiːtiəˈrɒlədʒi/ *n.* 气象学，气象状态

(14) oceanography /ˌəʊʃəˈnɒgrəfi/ *n.* 海洋学

(15) geophysics /ˌdʒiːəʊˈfɪzɪks/ *n.* 地球物理学

(16) geochemistry /dʒiːəʊˈkemɪstrɪ/ *n.* 地球化学

(17) stream bed：河床

(18) mica /ˈmaikə/ *n.* 云母

(19) field stone：原野石

(20) quartz crystal：石英晶体

Questions for review

(1) _____ is the study of the composition, structure, and history of the earth.

    A. Geology

    B. Paleontology

    C. Geophysics

    D. Geochemistry

(2) In their study of rocks and minerals geologists apply _____.

    A. chemical and physical concepts of atoms, molecules and radioactive isotopes

    B. chemical and physical concepts of atoms, molecules, and crystals

    C. findings on radioactive isotopes from atomic physics

    D. findings on atoms, molecules and radioactive isotopes

(3) _____ proves that the oceans have been hundreds of feet deeper than they are today.

    A. Waves from the earthquakes travel through the interior

    B. Chemical and X-ray

    C. Mapping of sands, travels, and boulders

    D. Examination of ancient caves and strand lines

(4) While geology has as its central themes _____, it is overlapped by many other earth-oriented sciences.

A. the study of the atmosphere

B. the materials and structure of the earth and the earth's history

C. the external features of the globe

D. the earth's waters and their uneven depths

(5) What is geology?

(6) What are the fundamental sciences on which geology depends?

(7) What can reveal the composition of rocks and minerals?

(8) Can you name some earth-oriented sciences? What are they?

Unit 2

Age of the Earth

Geologic Time

1 The earth is very old – 4.5 billion years or more – according to recent estimates. This vast **span** of time, called geologic time by the earth scientists, is difficult to comprehend in the familiar time units of months and years, or even centuries. How then do scientists **reckon** geologic time, and why do they believe the earth is so old? A great part of the secret of the earth's age is locked up in its rocks, and our centuries-old search for the key leads to the beginning and **nourishes** the growth of geologic science.

2 Mankind's **speculations** about the nature of the earth inspired much of the **lore** and legend of early civilizations, but at times there were only flashes of insight. The ancient historian Herodotus, in the 5th century B.C., made one of the earliest recorded geological observations. After finding fossil shells far inland in what are now parts of Egypt and Libya, he correctly inferred that the Mediterranean Sea had once extended much farther to the south. Few believed him, however, nor did the idea catch on[1]. In the 3rd century B.C., Eratosthenes **depicted** a **spherical** earth and even calculated its diameter and circumference, but the concept of a spherical earth was beyond the imagination of most people. Only 500

years ago, sailors aboard the Santa Maria[2] begged Columbus to turn back lest they sail off the earth's "edge"[3]. Similar opinions and prejudices about the nature and age of the earth have waxed and waned[4] through the centuries. Most people, however, appear to have traditionally believed the earth to be quite young – that its age might be measured in terms of thousands of years, but certainly not in millions.

3　The evidence for an ancient earth is **concealed** in the rocks that form the earth's **crust** and surface. The rocks are not all the same age but, like the pages in a long and complicated history, they record the earth-shaping events and life of the past. The record, however, is incomplete. Many pages, especially in the early parts, are missing and many others are **tattered,** torn, and difficult to **decipher**. But enough of the pages are preserved to reward readers with accounts of **astounding episodes** which **certify** that the earth is billions of years old.

4　Two scales are used to date these episodes and to measure the age of the earth: a relative time scale based on the sequence of layering of rocks and the evolution of life, and the **radiometric** time scale based on the natural **radioactivity** of chemical elements in some of the rocks[5]. An explanation of the relative scale **highlights** events in the growth of geologic science itself; the radiometric scale is a more recent development borrowed from physical sciences and applied to geologic problems[6].

5　At the close of the 18th century, the **haze** of fantasy and **mysticism** that tended to obscure the true nature of the earth was being swept away. Careful studies by scientists showed that rocks had diverse origins. Some rock layers, containing clearly identifiable fossil **remains** of fish and other forms of **aquatic** animal and plant life, originally formed in the ocean. Other layers, consisting of sand grains **winnowed** clean by the **pounding** surf, obviously formed as beach deposits that marked the shorelines of ancient seas[7]. Certain layers are in the form of sand bars[8] and **gravel** banks[9] – rock **debris** spread over the land by streams. Some rocks were once **lava** flows or beds of **cinders** and ash thrown out of ancient volcanoes; others are portions of large masses of once molten rock that cooled very slowly far beneath the earth's surface. Other rocks were so transformed by heat and pressure during the **heaving** and **buckling** of the earth's crust in periods of mountain building that their original features were **obliterated**.

6　Between the years of 1785 and 1800, James Hutton and William Smith advanced the concept of geologic time and strengthened the belief in an ancient world. Hutton, a Scottish geologist, first proposed formally the fundamental principle used to classify rocks according to their relative ages. He concluded, after studying rocks at many **outcrops**, that each layer represented a specific **interval** of geologic time. Further, he proposed that wherever **non-contorted** layers were exposed, the bottom layer was deposited first and was, therefore, the oldest layer deposited; each **succeeding** layer, up to the topmost one, was **progressively** younger.

7 Today, such a proposal appears to be quite elementary but, nearly 200 years ago, it amounted to[10] a major breakthrough in scientific reasoning by establishing a rational basis for relative time measurements. However, unlike tree-ring dating – in which each ring is a measure of 1 year's growth – no precise rate of deposition can be determined for most of the rock layers. Therefore, the actual length of geologic time represented by any given layer is usually unknown or, at best, a matter of opinion.

8 William Smith, a civil engineer and surveyor, was well acquainted with areas in southern England where "**limestone** and **shales** are layered like slices of bread and butter". His hobby of collecting and cataloging fossil shells from these rocks led to the discovery that certain layers contained fossils unlike those in other layers[11]. Using these key or **index** fossils as markers, Smith could identify a particular layer of rock wherever it was exposed. Because fossils actually record the slow but progressive development of life, scientists use them to identify rocks of the same age throughout the world.

9 From the results of studies on the origins of various kinds of rocks (**petrology**), coupled with studies of rock layering (**stratigraphy**) and the evolution of life (paleontology), geologists reconstruct the sequence of events that has shaped the earth's surface[12]. Their studies show, for example, that during a particular episode the land surface was raised in one part of the world to form high plateaus and mountain ranges. After the uplift of the land, the forces of **erosion** attacked the highlands and the eroded rock debris was transported and redeposited in the lowlands. During the same interval of time in another part of the world, the land surface **subsided** and was covered by the seas. With the sinking of the land surface, **sediments** were deposited on the ocean floor. The evidence for the pre-existence of ancient mountain ranges lies in the nature of the eroded rock debris, and the evidence of the seas' former presence is, in part, the fossil forms of marine life that accumulated with the bottom

sediments.

10 Such recurring events as mountain building and sea **encroachment**, of which the rocks themselves are records, comprise units of geologic time even though the actual dates of the events are unknown. Geologists have worked out the geologic time by dividing the earth's history into Eras based on the general character of life that existed during these times and Periods based on the evidence of major **disturbances** of the earth's crust.

11 The names used to **designate** the divisions of geologic time are a fascinating mixture of words that mark highlights in the historical development of geologic science over the past 200 years. Nearly every name **signifies** the acceptance of a new scientific concept – a new **rung** in the ladder of geologic knowledge.

New words

span /spæn/	*n.* 时距；跨度；间距；变化范围
	vt. 延续；横跨；贯穿；遍及
reckon /ˈrekən/	*v.* 计算；认为；估计
nourish /ˈnʌrɪʃ/	*v.* 滋养；给营养；培育；怀有
speculation /ˌspekjʊˈleɪʃn/	*n.* 推测；投机；沉思
lore /lɔː(r)/	*n.* 学问；知识；传说
depict /dɪˈpɪkt/	*vt.* 描述；描画
spherical /ˈsferɪkəl/	*adj.* 球的；球面的；球状的
conceal /kənˈsiːl/	*vt.* 隐藏；隐瞒；掩盖
tatter /ˈtætə(r)/	*vt.* 扯碎，撕碎
decipher /dɪˈsaɪfə(r)/	*n.* 密电译文
	vt. 破译（密码）；解释
astounding /əsˈtaʊndɪŋ/	*adj.* 令人震惊的；令人惊骇的
episode /ˈepɪsəud/	*n.* 一段经历；插曲；片段
certify /ˈsɜːtɪfaɪ/	*vt.* 证明；发证书给……
radiometric /ˌreɪdɪəuˈmetrɪk/	*adj.* 辐射度测量的，放射性测量的
radioactivity /ˌreɪdɪəuækˈtɪvətɪː/	*n.* 放射性；辐射
highlight /ˈhaɪlaɪt/	*vt.* 强调；照亮；使突出
	n. 加亮区；精彩部分；最重要的细节或事件；

闪光点

haze /heɪz/	n. 雾团，霾
mysticism /ˈmɪstəˌsɪzəm/	n. 神秘；神秘主义；谬论
remains /rɪˈmeɪnz/	n. 遗迹；遗体
aquatic /əˈkwætik/	adj. 水生的；含水的
	n. 水生动植物
winnow /ˈwɪnəu/	n. 扬谷；扬谷器
	vt. 簸；把……挑出来；精选
	vi. 分出好坏；扬谷
pound /paund/	n. 磅；英镑
	vt. 捣碎；敲打；连续砰砰地猛击
	vi. 咚咚地走；（心脏）怦怦地跳；砰砰地敲
gravel /ˈgræv(ə)l/	n. 沙砾，碎石
debris /ˈdeɪbrɪ/	n. 岩屑，碎石，尾矿
lava /ˈlɑːvə/	n. 熔岩；火山岩
cinder /ˈsɪndə(r)/	n. 火山渣；炉渣
heaving /ˈhiːvɪŋ/	n. 冻胀，隆起
buckling /ˈbʌklɪŋ/	n. 弯折，挤弯作用
obliterate /əˈblɪtəreɪt/	vt. 除去，消灭
outcrop /ˈautkrɒp/	n. 露头；露出地面的岩层
	vi. 露出
contorted /kənˈtɔːtid/	adj. 弯曲的，扭曲的
	v. 扭曲；歪曲（contort 的过去分词）
succeeding /səkˈsiːdɪŋ/	adj. 随后的，以后的
progressively /prəˈgresivli/	adj. 渐进地，日益增加地
limestone /ˈlaɪmstəun/	n. 石灰岩
shale /ʃeɪl/	n. [岩] 页岩
index /ˈɪndeks/	n. 指标；指数；索引；指针
	vt. 指出；编入索引
	vi. 做索引
petrology /pəˈtrɒlədʒɪ/	n. 岩石学

stratigraphy /strəˈtɪɡrəfɪ/　　　　　*n.* 地层学

erosion /ɪˈrəʊʒn/　　　　　*n.* 侵蚀，冲刷

subside /səbˈsaɪd/　　　　　*vi.* 减弱，（热度）消退；沉淀

sediment /ˈsedɪmənt/　　　　　*n.* 沉积；沉淀物

recur /rɪˈkɜː(r)/　　　　　*vi.* 再发生，复发

encroachment /ɪnˈkrəʊtʃmənt/　　　　　*n.* 侵入，侵蚀

disturbance /dɪˈstɜːbəns/　　　　　*n.* 困扰；动乱；干扰；妨碍

designate /ˈdezɪɡneɪt/　　　　　*vt.* 指明，指出；指派

signify /ˈsɪɡnɪfaɪ/　　　　　*vt.* 表示；意味（着）

rung /rʌŋ/　　　　　*n.* 梯级，阶梯；地位

　　　　　v. 打电话（ring 的过去式和过去分词）

Notes to the text

(1) catch on：理解；变得流行

(2) the Santa Maria：圣马利亚号（哥伦布第一次远航发现美洲所乘之船）

(3) Only 500 years ago, sailors aboard the Santa Maria begged Columbus to turn back lest they sail off the earth's "edge".

　　lest（*conj.* 唯恐；以免；担心）后接（should）do 的虚拟语气结构。

译文：在 500 年前，圣马利亚号上的水手还因为担心到了地球的"边缘"而请求哥伦布返航。

(4) wax and wane：盈亏；盛衰

(5) Two scales are used to date these episodes and to measure the age of the earth: a relative time scale based on the sequence of layering of rocks and the evolution of life, and the radiometric time scale based on the natural radioactivity of chemical elements in some of the rocks.

　　本句中的两个 based on 均是过去分词做后置定语，分别修饰 a relative time scale 和 the radiometric time scale。

译文：可以用两种时间尺度来测定地球年龄和确定地质事件的形成时间：根据岩层层序来确定的相对年代和根据岩石中化学元素的天然放射性进行放射性测年的绝对年代。

(6) The radiometric scale is a more recent development borrowed from the physical sciences and applied to geologic problems.

borrowed... 和 applied... 是过去分词结构做后置定语，修饰 a more recent development。

译文：同位素地质年代法是借助于物理学科并应用于地质问题研究的一种较新的方法。

(7) Other layers, consisting of sand grains winnowed clean by the pounding surf, obviously formed as beach deposits that marked the shorelines of ancient seas.

本句的基本结构是 Other layers obviously formed as beach deposits。consisting of ... 是现在分词做后置定语，修饰 other layers；winnowed clean...是过去分词做后置定语，修饰 sand grains；that marked...是定语从句，修饰 beach deposits。

译文：其他岩层沉积在海滨，其含有的沙粒被拍岸海浪冲刷干净，清晰地标示着古代海洋的海岸线。

(8) sand bars：河口沙洲；沙坝

(9) gravel banks：砾石岸；砾滩

(10) amount to：总计；相当于

(11) His hobby of collecting and cataloging fossil shells from these rocks led to the discovery that certain layers contained fossils unlike those in other layers.

that certain layers...是同位语从句，说明 discovery 的内容。

译文：他喜欢从岩石上收集贝壳化石并把它们逐一编录。这个爱好使他发现某些岩层中的化石与其他岩层中的化石不一样。

(12) From the results of studies on the origins of various kinds of rocks (petrology), coupled with studies of rock layering (stratigraphy) and the evolution of life (paleontology), geologists reconstruct the sequence of events that has shaped the earth's surface.

coupled with（与……联合/结合）是过去分词短语（被动语态）做 the results of studies 的后置定语。

译文：地质学家们将岩石学的研究结果与地层学和古生物学的研究结合，重现了塑造地表的地质事件的先后顺序。

Understand the text

Answer the following questions according to the passage you have read.

(1) What is geologic time?

(2) What did Herodotus find in the 5th century B.C.?

(3) What does the story of the sailors aboard the Santa Maria begging Columbus to turn back show?

(4) What roles do the rocks play when it comes to the age of the earth?

(5) What are the differences of the two scales in measuring the age of the earth?

(6) What is James Hutton's contribution to the measuring of the geologic time?

(7) What did William Smith discover?

(8) What is the significance of the studies on the origins of the rocks, rock layering and the evolution of life?

(9) What has led to the formation of the Earth's surface as we know today according to the passage?

(10) What have geologists done with the geologic time?

Translation

1. Translate the following sentences into Chinese.

(1) A great part of the secret of the earth's age is locked up in its rocks, and our centuries-old search for the key led to the beginning and nourished the growth of geologic science.

_____

_____

_____

(2) Some rock layers, containing clearly identifiable fossil remains of fish and other forms of aquatic animal and plant life, originally formed in the ocean.

_____

_____

_____

(3) Wherever non-contorted layers were exposed, the bottom layer was deposited first and was, therefore, the oldest layer deposited.

_____

_____

(4) Because fossils actually record the slow but progressive development of life, scientists use them to identify rocks of the same age throughout the world.

_____

_____

(5) Such recurring events as mountain building and sea encroachment, of which the rocks themselves are records, comprise units of geologic time even though the actual dates of the events are unknown.

_____

_____

_____

2. Translate the following passage into English.

地质年代是用来描述地球历史事件的时间单位和各种地质事件发生的时间。它包含两方面的含义：其一是指各地质事件发生的时间先后顺序，称为相对地质年代；其二是指各地质事件从发生到现今的时间段，称为绝对地质年代。这两方面结合，才构成对地质事件及地球、地壳演变时代的完整认识，地质年代表正是在此基础上建立起来的。

_____

_____

_____

_____

_____

_____

_____

_____

_____

_____

_____

_____

_____

_____

_____

_____

Writing skill

Features of EST at Syntactic Level

To achieve clarity, logicality and objectivity, the EST is of typical features at syntactic level, which cover *passive voice, modifiers, non-finite verbs and complex sentences.*

Passive voice is frequently used in EST to make the sentences sound more objective instead of subjective. For example:

The evidence for an ancient earth is concealed in the rocks that form the earth's crust and surface. (Para.3)

If this sentence is written this way: "We concealed the evidence for an ancient earth in the rocks that form the earth's crust and surface", the readers may question that it is "we" who subjectively concealed the evidence. Hence, the passive voice here used without referring to some certain people indicates the objectivity of the findings.

Modifiers, including pre- modifiers and post- modifiers, are used to make sentences clear and concise, especially in definition. For example:

The radiometric scale is a more recent development borrowed from physical sciences and applied to geologic problems.(Para.4)

In this sentence, "radiometric" is a pre-modifier to indicate the type of scale; while "borrowed from physical sciences" and "applied to geologic problem" are post-modifiers to further explain the "development".

Non-finite verbs have no tense, person, or singular/plural form. The infinitive, present and past participles are the non-finite parts of a verb: to do; doing; done. In EST, non-finite verbs are used to replace adverbial clause or attributive clause.

The following are some more examples taken from Text A:

Two scales are used to date these episodes and to measure the age of the earth.(Para.4)

The underlined parts are two infinitives showing the purposes of using the two scales.

Some rock layers, containing clearly identifiable fossil remains of fish and other forms of aquatic animal and plant life, originally formed in the ocean.(Para.5)

"containing", a present participle here, functions as attribute to limit "some rock layers".

Therefore, the actual length of geologic time represented by any given layer is usually unknown or, at best, a matter of opinion.(Para.7)

"represented" is a past participle used to modify "geologic time".

Complex sentences are long sentences with complex components (phrases, clauses or modifiers) commonly used in EST to express complicated ideas clearly, such as sentences with attributive clause, adverbial clause or noun clause, etc.

Some examples are taken from Text A as follows:

The evidence for an ancient earth is concealed in the rocks that form the earth's crust and surface.(Para.3)

The underlined part is a typical attributive clause modifying "the rocks" before it.

> *Careful studies by scientists showed <u>that rocks had diverse origins.</u> (Para.5)*
>
> The underlined part acts as an object clause of "showed".

Writing practice

You may try to pick out more examples with syntactical features mentioned above of EST from both Text A and Text B.

Text B

Geologic Time Scale

1 The Geologic Time Scale is the history of the earth broken down into spans of time marked by various geologic events. There are other markers, like the types of species and how they evolved, that distinguish one time from another on the Geologic Time Scale.

2 There are four main time spans that generally mark the Geologic Time Scale divisions. The first, Precambrian[1] Time is not an actual era on the Geologic Time Scale because of the lack of diversity of life, but the other three divisions are defined eras. The Paleozoic[2] Era, Mesozoic[3] Era and Cenozoic[4] Era saw many great changes (Figure 2-1) .

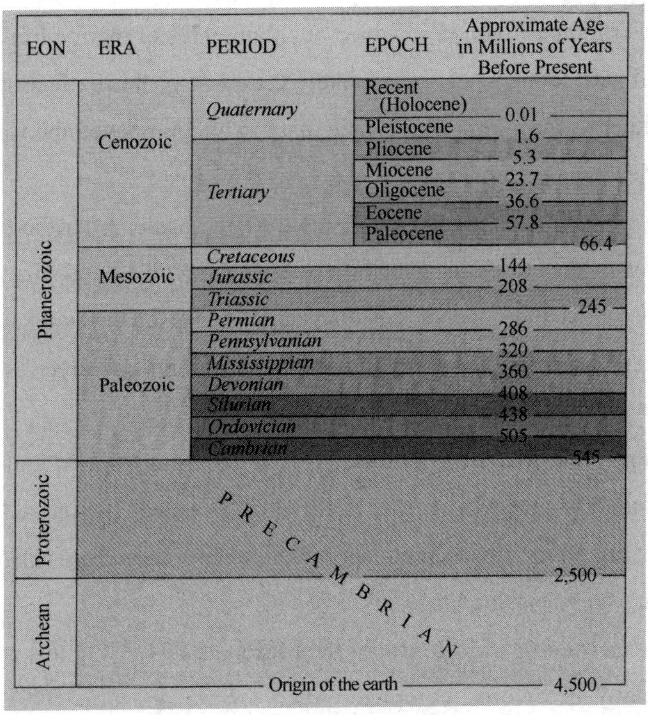

EON	ERA	PERIOD	EPOCH	Approximate Age in Millions of Years Before Present
Phanerozoic	Cenozoic	Quaternary	Recent (Holocene)	0.01
			Pleistocene	1.6
		Tertiary	Pliocene	5.3
			Miocene	23.7
			Oligocene	36.6
			Eocene	57.8
			Paleocene	66.4
	Mesozoic	Cretaceous		144
		Jurassic		208
		Triassic		245
	Paleozoic	Permian		286
		Pennsylvanian		320
		Mississippian		360
		Devonian		408
		Silurian		438
		Ordovician		505
		Cambrian		545
Proterozoic		PRECAMBRIAN		2,500
Archean				
		Origin of the earth		4,500

Figure 2-1   The Geological Time Scale

3 The Precambrian Time Span began at the beginning of the earth 4.6 billion years ago. For billions of years there was no life on the earth. It wasn't until the end of this time period

that single celled organisms came into existence. No one knows for sure how life on the earth began, but there are several theories like the Primordial Soup Theory[5], Hydrothermal Vent Theory[6], and Panspermia Theory[7].

4　The end of this time span saw the rise of a few more complex animals in the oceans like jellyfish. There was still no life on land and the atmosphere was just beginning to accumulate the oxygen needed for higher order animals to survive. It wasn't until the next era that life really began to take off and diversify.

5　The Paleozoic Era began with the Cambrian Explosion[8]. This relatively rapid period of large amounts of speciation kicked off[9] a long time span of flourishing life on the earth. These great amounts of life in the oceans soon moved onto land. First, plants made the move and then invertebrates[10]. Not long after that, vertebrates moved to land as well. Many new species appeared and thrived.

6　The end of the Paleozoic Era came with the largest mass extinction in the history of life on the earth. The Permian Extinction[11] wiped out about 95% of marine life and nearly 70% of life on land. Climate changes were most likely the cause of this extinction as the continents all drifted together to form Pangaea[12]. The mass extinction paved the way for new species to arise and a new era to begin.

7　The Mesozoic Era is the next era on the Geologic Time Scale. After the Permian Extinction caused so many species to extinct, many new species evolved and thrived. The Mesozoic Era is also known as the "age of the dinosaurs" because dinosaurs were the dominant species for much of the era. Dinosaurs started off small and got larger as the Mesozoic Era went on.

8　The climate during the Mesozoic Era was very humid and tropical and many lush, green plants were found all over the earth. Herbivores[13] especially thrived during this time period. Besides dinosaurs, small mammals came into existence. Birds also evolved from the dinosaurs during the Mesozoic Era.

9　Another mass extinction marks the end of the Mesozoic Era. All dinosaurs, and many other animals, especially herbivores, completely died off. Again, niches[14] were needed to be filled by new species in the next era.

10　The last and current time period on the Geologic Time Scale is the Cenozoic Period. With large dinosaurs' extinction, the smaller mammals that survived were able to grow and become dominant life on the earth. Human evolution also all happened during the Cenozoic Era.

11 The climate has changed drastically over the relatively short amount of time in this period. It got much cooler and drier than the Mesozoic Era climate. There was an ice age where most temperate parts of the earth were covered with glaciers, which made life have to adapt rather rapidly and increased the rate of evolution.

12 All life on the earth evolved into their present day forms. The Cenozoic Era has not ended and most likely will not end until another mass extinction period.

Notes to the text

(1) Precambrian /priˈkæmbriən/ n. 前寒武纪

(2) Paleozoic /ˌpælɪəuˈzəʊɪk; -ˌpei-/ n. 古生代

(3) Mesozoic /ˌmesəuˈzəʊɪk; ˌmez-/ n. 中生代

(4) Cenozoic /ˌsiːnəˈzəʊɪk/ n.（等于 Cainozoic）新生代

(5) Primordial Soup Theory：原生浆液说

(6) Hydrothermal Vent Theory：海底热泉说

(7) Panspermia Theory：胚种论

(8) Cambrian Explosion：寒武纪大爆发

(9) kick off：开始

(10) invertebrate /ɪnˈvɜːtɪbrət, -breɪt/ n. 无脊椎动物

(11) Permian Extinction：二叠纪灭绝

(12) pangaea /pænˈdʒiːə/ n. 泛古陆；泛大陆

(13) herbivore /ˈhɜːbɪvɔː/ n. 草食动物

(14) niche /niːʃ; nɪtʃ/ n.［生］生态龛；生态位；小生态环境

Questions for review

(1) The Geologic Time Scale _____.

    A. marks spans of time by various events

    B. distinguishes one time from another by the types of species

    C. shows the history of the earth from different markers

    D. divides the history of the earth into eras

(2) During the Precambrian Time, _____.

    A. life on the earth began to diversify

    B. the environment was good only for single celled organisms

    C. there was no life on the earth probably because of the lack of oxygen in the atmosphere

    D. the oceans were better places for complex animals

(3) From what happened in the Paleozoic Era, we may infer that _____.

    A. there was no life left on the earth because of the mass extinction

    B. life on the earth evolved step by step

    C. the climate was hot enough to wipe out most of the life on the earth

    D. the mass extinction caused the climate changes

(4) Which of the following is NOT true according to the passage?

    A. The Mesozoic Era is also called the "age of the dinosaurs" because dinosaurs were the dominant species of the time.

    B. The climate during the Mesozoic Era was responsible for the extinction of dinosaurs.

    C. The Cenozoic Era saw the evolution of human beings.

    D. The Cenozoic Era is the period we are living in now.

(5) This passage is mainly about _____.

    A. the origin of life on the earth

    B. the climate changes on the earth

    C. the geologic time divisions of the earth

    D. the evolution of life on the earth

(6) What geologic time divisions has the author described? Why do geologists divide the history of the earth in this way?

(7) What features are there in different geologic periods respectively?

(8) What can you expect the earth to be in the future?

Unit 3

Exodynamic Geology

Weathering and Erosion

1 Rocks exposed on the earth's surface are constantly being altered by water, air, changing temperature, and other environmental factors. The term "weathering" refers to the group of destructive processes that change the physical and chemical character of rock on or near the earth's surface. The tightly bound crystals of an igneous rock[1] can be loosened and altered to new minerals by weathering. Weathering can be a mechanical or a chemical process. Often, these two types of weathering work together.

Mechanical Weathering

2 Mechanical or physical weathering involves the breakdown of rocks and soils through direct contact with atmospheric conditions such as heat, water, ice and pressure.

3 Water seeps into cracks and **crevices** in rock. If the temperature drops low enough, the water will freeze. When water freezes, it expands. The ice then works as a **wedge**. It slowly widens the cracks and splits the rock. When ice melts, water performs the act of erosion by carrying away the tiny rock fragments lost in the split[2].

4　Mechanical weathering also occurs as rock heats up and cools down. The changes in temperature cause rock to expand and contract. As this happens over and over again, the rock weakens. Over time, it **crumbles**.

5　Another type of mechanical weathering occurs when clay or other materials near hard rock absorb water. The clay swells with the water, breaking apart the surrounding rock[3] (Figure 3-1) .

Figure 3-1　Spheroidal Weathering

6　Salt also works to weather rock. Saltwater sometimes gets into the cracks and pores of rock. If the saltwater **evaporates**, salt crystals are left behind. As the crystals grow, they put pressure on the rock, slowly breaking it apart.

Chemical Weathering

7　The second classification, chemical weathering, involves the **decomposition** of rock from exposure to water and atmospheric gases (principally carbon dioxide and water vapor). As rock is **decomposed** by these agents, new chemical compounds form. Chemical weathering changes the materials that make up rocks and soil. Sometimes, carbon dioxide from air or soil combines with water. This produces a weak acid called carbonic acid[4], which can dissolve rock.

8 Carbonic acid is especially effective at dissolving limestone. When the carbonic acid seeps through limestone underground, it can open up huge cracks or hollow out[5] vast networks of caves. Carlsbad Caverns National Park[6], in the State of New Mexico, U.S.A, includes more than 110 limestone caves. The largest is called the Big Room. With about 1,200 meters long and 190 meters wide, it is the size of six football fields.

9 Sometimes, chemical weathering dissolves large regions of limestone or other rocks on the surface of the earth to form a landscape called **karst**. In these dramatic areas, the surface rock is **pockmarked** with holes, **sinkholes**, and caves. One of the world's most spectacular examples of karst is Shilin, or the Stone Forest, near Kunming, China. Hundreds of slender, sharp towers of limestone rise from the landscape.

10 Another type of chemical weathering works on rocks that contain iron. These rocks rust in a process called **oxidation**. As the rust expands, it weakens the rock and helps break it apart.

11 Erosion is the process by which soil and rock are removed from the earth's surface by **exogenetic** processes such as wind or water flow, and then transported and deposited in other locations.

Erosion by Water

12 Moving water is the major agent of erosion. Rain carries away bits of soil and slowly washes away rock fragments. Rushing streams and rivers wear away[7] their banks, creating larger and larger valleys. In a span of about 5 million years, the Colorado River[8] cut deeper and deeper into the land in what is now the U.S. state of Arizona. It eventually formed the Grand Canyon[9], which is more than 1,600 meters deep and as much as 29 kilometers wide in some places.

13 Erosion by water changes the shape of coastlines. Waves constantly crash against shores. They pound rocks into **pebbles** and **reduce** pebbles to sand. Water sometimes takes sand away from beaches. This moves the coastline farther inland.

14 The battering of ocean waves also erodes seaside cliffs. It sometimes bores holes that form caves. When water breaks through the back of the cave, it creates an **arch**. The continual pounding of waves can cause the top of the arch to fall, leaving nothing but rock columns. These are called sea stacks[10]. All of these features make rocky beaches beautiful, but also

dangerous.

Erosion by Wind

15　Wind is also an agent of erosion. It carries dust, sand, and volcanic ash from one place to another. Wind can sometimes blow sand into **towering dunes**. Some sand dunes in some areas of the Gobi Desert[11] in China reach more than 400 meters high.

16　In dry areas, windblown sand blasts against rock with tremendous force, slowly wearing away the soft rock. It also polishes rocks and cliffs until they are smooth.

17　Wind is responsible for the dramatic arches that give Arches National Park[12], in the U.S. state of Utah, its name. Wind can also erode material until nothing remains at all. Over millions of years, wind and water eroded an entire mountain range in central Australia. Uluru[13], also known as Ayers Rock[14], is the only **remnant** of those mountains.

Erosion by Ice

18　Ice can erode land. In **frigid** areas and on some mountaintops, glaciers move slowly downhill and across the land. As they move, they pick up everything in their path, from tiny grains of sand to huge boulders.

19　The rocks carried by a glacier rub against the ground below, eroding both the ground and the rocks. Glaciers grind up[15] rocks and scrape away the soil. Moving glaciers gouge out[16] basins and form steep-sided mountain valleys.

20　Today, in places such as Greenland and Antarctica, glaciers continue to erode the earth. These ice sheets, sometimes more than a mile thick, carry rocks and other debris downhill toward the sea. Eroded sediment is often visible on and around glaciers. This material is called **moraine**.

21　It is important to distinguish between weathering and erosion. Weathering breaks down rocks that are either stationary or moving. Erosion is the picking up or physical removal of rock particles by an agent such as streams, wind or glaciers. Weathering helps break down a solid rock into loose particles that are easily eroded. Most eroded rock particles are at least partially weathered, but rock can be eroded before it has been weathered at all. A stream can erode weathered or unweathered rock fragments. Far more erosion occurs naturally, and a combination of weathering and erosion is responsible for producing the soil from which earth's plants grow.

22 Weathering and erosion slowly **chisel**, polish, and **buff** the earth's rock into ever evolving works of art – and then wash the remains into the sea.

23 Working together, they create and reveal marvels of nature from tumbling boulders high in the mountains to sandstone arches in the **parched** desert to polished cliffs braced against violent seas.

New words

crevice /ˈkrevɪs/　　　　　　　*n.* 裂缝；裂隙

wedge /wedʒ/　　　　　　　*n.* 楔形物

crumble /ˈkrʌmbl/　　　　　　*vi.* 崩溃；破碎，崩解

　　　　　　　　　　　　　vt. 崩溃；弄碎，粉碎

evaporate /ɪˈvæpəreɪt/　　　　*vt.* 蒸发；使……脱水；使……消失

　　　　　　　　　　　　　vi. 蒸发，挥发；消失，失踪

decomposition /ˌdiːkɒmpəˈzɪʃən/　*n.* 分解，腐烂；变质

decomposed /ˌdiːkəmˈpəʊzd/　*adj.* 已腐烂的，已分解的

karst /kɑːst/　　　　　　　*n.* 喀斯特（石灰岩地区常见的地形）；岩溶

pockmark /ˈpɒkmɑːk/　　　　*n.* 麻子；凹坑

　　　　　　　　　　　　　vt. 使留下痘疤；使有凹坑

sinkhole /ˈsɪŋkhəʊl/　　　　*n.* 落水洞；灰岩坑

oxidation /ˌɒksɪˈdeɪʃn/　　　*n.* [化] 氧化

exogenetic /ˈeksəʊdʒɪˈnetɪk/　*adj.* 外生的；外因的；外源性的

pebble /ˈpebl/　　　　　　　*n.* 砾石，鹅卵石

　　　　　　　　　　　　　v.（用卵石等）铺

reduce /rɪˈdjuːs/　　　　　　*vt.* 缩减；简化；还原

arch /ɑːtʃ/　　　　　　　　*n.* 弓形，拱形；拱门

　　　　　　　　　　　　　vt. 使……弯成弓形；用拱连接

　　　　　　　　　　　　　vi. 拱起；成为弓形

towering /ˈtaʊərɪŋ/　　　　　*adj.* 高耸的；卓越的；激烈的

dune /djuːn/　　　　　　　*n.*（由风吹积而成的）沙丘

remnant /ˈremnənt/　　　　　*n.* 剩余部分

	adj. 剩余的
frigid /ˈfrɪdʒɪd/	*adj.* 寒冷的，严寒的
moraine /məˈrein/	*n.* 冰碛；（熔岩流表面的）火山碎屑
chisel /ˈtʃɪzl/	*vt.* 雕，刻；凿；欺骗
	vi. 雕，刻；凿；欺骗
	n. 凿子
buff /bʌf/	*vt.* 软皮摩擦；缓冲；擦亮，抛光某物
parched /pɑːtʃt/	*adj.* 焦的；炎热的；炒过的；干透的
	vt. 烘干；使极渴（parch 的过去分词）

Notes to the text

(1) igneous rock：火成岩

(2) When ice melts, water performs the act of erosion by carrying away the tiny rock fragments lost in the split.

译文：当冰融化时，水便带走留在裂缝中的细小岩石碎片完成侵蚀作用。

(3) surrounding rock：围岩

(4) carbonic acid：碳酸

(5) hollow out：挖空

(6) Carlsbad Caverns National Park：（美）卡尔斯巴德洞窟国家公园

(7) wear away：磨损；消磨；流逝

(8) the Colorado River：（美）科罗拉多河

(9) the Grand Canyon：（美）大峡谷

(10) sea stacks：海蚀柱；海柱

(11) the Gobi Desert：戈壁沙漠

(12) Arches National Park：（美）拱门国家公园

(13) Uluru：乌卢鲁（澳大利亚艾尔斯岩，世界最大的单体巨石）

(14) Ayers Rock：（澳）艾尔斯巨石

(15) grind up：磨碎

(16) gouge out：挖出；凿槽

Understand the text

Answer the following questions according to the passage you have read.

(1) What is the definition of weathering?

(2) How do atmospheric factors affect the process of mechanical weathering?

(3) In what ways does chemical weathering work on rocks?

(4) How many agents of erosion are mentioned here? What are they?

(5) How does erosion by water change the shape of coastlines?

(6) What forms sand dunes in some areas of the Gobi Desert?

(7) How can ice erode the land?

(8) How do you distinguish between weathering and erosion?

Translation

1. Translate the following sentences into Chinese.

(1) The tightly bound crystals of an igneous rock can be loosened and altered to new minerals by weathering.

_____

_____

_____

_____

_____

_____

(2) The changes in temperature cause rock to expand and contract. As this happens over and over again, the rock weakens. Over time, it crumbles.

_____

_____

_____

_____

_____

(3) Carbonic acid is especially effective at dissolving limestone. When the carbonic acid seeps through limestone underground, it can open up huge cracks or hollow out vast networks of caves.

_____

_____

_____

_____

(4) The battering of ocean waves also erodes seaside cliffs. It sometimes bores holes that form caves.

_____

_____

_____

_____

(5) Weathering and erosion slowly chisel, polish, and buff the earth's rock into ever evolving works of art–and then wash the remains into the sea.

_____

_____

_____

_____

2. Translate the following passage into English.

全球变暖，世界各地的温度升高，正在加快侵蚀的速度。气候的变化也与更频繁、更剧烈的风暴联系起来。伴随着飓风与台风后的风暴潮可能侵蚀数英里的海岸线和沿海栖息地。这些沿海地区有家园、企业，以及经济上重要的产业，比如渔场。温度的升高也迅速融化着冰川，这就造成了海平面上升的速度超出了生物可以适应的程度。上升的海平面会更迅速地侵蚀沙滩。据估计，海平面上升 8～10 厘米引起的侵蚀将足以威胁到建筑物、下水道系统、道路和隧道。

_____

_____

_____

_____

_____

_____

_____

_____

_____

_____

Writing skill

Steps of Choosing a Topic

A topic is the main organizing principle guiding the analysis of a research paper. Topics offer an occasion for writing and a focus that governs what the author intends to express. Topics represent the core subject matter of scholarly communication. Be aware that selecting a good topic may not be easy. It must be narrow and focused enough to be probed into, yet broad enough to find adequate information to extend. The following four steps are suggested for choosing an appropriate topic.

1. Choose something that holds your interest

If you feel strongly interested in a topic, you will write with more enthusiasm. View the topic from a variety of perspectives and find the angle that interests you most. Readers respond positively to sincere interest conveyed in words.

2. Think of the 5W questions

WHY do you choose the topic? Do you have an opinion about the issues involved?

WHO are the information providers on this topic? Who is affected by the topic? Do you know of organizations or institutions affiliated with the topic?

WHAT are the major questions for this topic? Are there a range of issues and viewpoints to consider?

WHERE is your topic important: at the local, national or international level? Are there specific places affected by the topic?

WHEN is/was your topic important? Is it a current event or an historical issue?

3. Analyze the chosen topic

A manageable topic is a topic that isn't too broad or too narrow. If the topic is too broad, you need to narrow down the scope of topic and make it specific. Many of the research topics you find through internet search are too general and could involve many different aspects. For example, "Weathering and Erosion", the topic of Text A in Unit 3 is an appropriate topic for an expository writing about popular science, but it may be an extremely broad topic for a study of academic writing. It is unclear whether you need to write about the phenomena of weathering and erosion around the world, or you should focus on the phenomena of weathering and erosion in a particular country, or you attach much attention to the effects of weathering and erosion bring about on geologic formation. You also need to indicate the domain in which the two phenomena you are demonstrating fall. Therefore, you may narrow down your topic as "A Study of Weathering and Erosion of Rock Slope in Chengdu". It is narrow enough, but this topic is also problematic since the addition of a location to this topic may result in very few results, which means the topic is too narrow to extend further. Depending on the type of research you are conducting, you may focus your study on a larger geographic limit. A better choice would be "A Study of Weathering and Erosion of Rock Slope in Sichuan Basin".

4. Read and research more about your topic

Once you have chosen a topic, find relevant information about it. You can browse through some current journals in your subject discipline. You only need one to be the spark that begins the process of wanting to learn more about a topic. List the key words describing your topic and do the search. Summarize the background information about your topic and note down the important people and the important works in the field. When all your criteria fail, go back and generate other topics. Don't give up easily. There are lots of topics within your niche. In addition, make sure there are adequate sources available on the topic. Even when your topic is appropriate, if there are no materials on it, it will be impossible to write about.

Writing practice

You may try to find out some research topics based on Text A.

Text B

Fluvial Process

1 Fluvial[1] processes comprise erosion and transportation of sediment or deposition on river bed.

2 Fluvial erosion is the removal of rock and other mineral particles from channel beds[2] and banks by stream flow. There are various types of erosion. Abrasion[3] is the grinding effect of a channel caused by sediments in the process of being transported. The material being transported undermines[4] river banks and valley slopes. However, the sediment itself is eroded by colliding with other pieces of material, which is known as attrition[5]. Hydraulic action[6] is the erosive effect of flowing water without assistance of rock particles. The sheer force of the flow causes erosion to occur. This type of erosion is most effective in areas of channels which consist of incoherent materials, such as sand and gravel. Cavitations[7] occur when tiny bubbles of air implode[8] in fissures[9] and cracks in channel banks. The tiny shock waves result in the weakening of river banks and the collapse of landforms. Corrosion[10] is a chemical action of stream water, which dissolves carbonate rocks such as chalk and limestone.

3 A river has three main stages: sediment supply zone, sediment transport zone and sediment storage zone. Within these three stages various erosions occur at each stage.

4 Within sediment supply zone, erosion is mainly directed vertically and headward, which is due to the river not having a lot of spare energy as it is using 90% of its energy to overcome obstacles such as large rocks and boulders. This erosion leads landforms such as rapids[11], small waterfalls and steep river bed to form (Figure 3-2).

5 Within sediment transport zone, it is mostly corrosion and attrition that occur due to the sediment being transported and colliding with both the channel bed and with each other. However, there is little hydraulic power and corrosion still taking place at this stage. Due to the river having fewer obstacles to overcome, it has more spare energy to erode more laterally rather than vertically, widening the river channel. This erosion leads to the

development of landforms such as rapids, small meanders[12], small floodplains[13], pools and riffles[14].

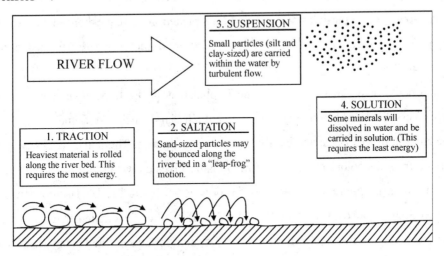

Figure 3-2　River Processes: Transportation

6　In sediment storage zone, less even no erosion occurs besides some lateral erosion[15] on the outside bends of meanders. This leads to the development of larger meanders and floodplains.

7　Solid and soluble particles eroded from channel, together with materials input by mass movements and weathering from valley slopes, are transported down the stream (from the upper stage to the lower stage). The material transported by a river is carried as either bedload[16] – the coarser particles that move near to river bed or suspended load[17] – the smaller particles which are carried in water. Another material which is carried by river in fluvial processes is dissolved material.

8　Material may be transported by a river in four main ways: solution, suspension, saltation[18] and traction[19].

9　The type of transport taking place depends on the size of sediment and the amount of energy that is available to undertake the transport. In the upper course of the river there is more traction and saltation going on due to the large size of the bedload; as the river enters its middle and lower course there is a lot of finer material eroded from further upstream which will be carried in suspension.

10　River deposition is where the material carried by river is dropped. A river is continually picking up and dropping solid particles of rock and soil from its bed throughout its length. When the river flows fast, more particles are picked up than dropped. When the river flows slowly, more particles are dropped than picked up. Areas where more particles are dropped are called alluvial[20] or flood plains, and the dropped particles are called alluvium[21].

11　Even small streams make alluvial deposits, but it is in the flood plains and deltas[22] of large rivers that large, geologically-significant alluvial deposits are found.

12　The amount of matter carried by a large river is enormous. The names of many rivers are derived from the color that the transported matter gives the water. For example, the Huang He in China is literally translated as "Yellow River", and the Mississippi River in the United States is also called "the Big Muddy." It has been estimated that the Mississippi River annually carries 406 million tons of sediment to the sea, the Huang He 796 million tons, and the Po River in Italy 67 million tons.

Notes to the text

(1) fluvial /ˈfluːvɪəl/ *adj.* 河流的；冲积的

(2) channel bed：河床

(3) abrasion /əˈbreɪʒn/ *n.* 磨蚀，海蚀，浪蚀，研磨

(4) undermine /ˌʌndəˈmaɪn/ *vt.* 渐渐破坏

(5) attrition /əˈtrɪʃn/ *n.* 摩擦；磨损

(6) hydraulic action：水力作用

(7) cavitation /ˌkævɪˈteɪʃ(ə)n/ *n.* 成穴；空化作用；气蚀；空穴现象

(8) implode /ɪmˈpləʊd/ *vt.* 向内破裂；内爆；突然崩溃；向心聚爆；向内坍塌

(9) fissure /ˈfɪʃə(r)/ *n.* 裂缝；裂沟（尤指岩石上的）

(10) corrosion /kəˈrəʊʒn/ *n.* 腐蚀

(11) rapids /ˈræpɪdz/ *n.* ［水文］急流；湍流

(12) meander /miˈændə(r)/ *n.* 曲流（常用复数）；河曲

(13) floodplain /ˈflʌdpleɪn/ *n.* 泛滥平原，漫滩，洪积平原

(14) riffle /ˈrɪfl/ *n.* 浅滩

(15) lateral erosion：侧蚀；旁蚀

(16) bedload /ˈbedləud/ *n.* 推移质

(17) suspended load：悬移质

(18) saltation /sælˈteɪʃ(ə)n; sɔː-; sɒ-/ *n.* 突变；（水中砂粒）跃移；不连续变异

(19) traction /ˈtrækʃn/ *n.* 拖拉，牵引；拉曳

(20) alluvial /əˈluːvɪəl/ *adj.* 冲积的

(21) alluvium /əˈluːvɪəm/ *n.* ［地质］冲积层，冲积土

(22) delta /ˈdeltə/ *n.*（河流的）三角洲

Questions for review

(1) The sediment itself is eroded by colliding with other pieces of material, which is known as _____.

　　A. hydraulic action

　　B. cavitations

　　C. corrosion

　　D. attrition

(2) Within the sediment transport zone, the river has more spare energy to erode more _____.

　　A. vertically

　　B. laterally

　　C. headward

　　D. backward

(3) In the sediment storage zone, which one of the following landforms does the erosion lead to?

　　A. Rapids.

　　B. Small waterfalls.

　　C. Larger meanders.

　　D. Pools and riffles.

(4) The _____ material transported by a river is called bedload.

　　A. coarser

　　B. smaller

C. dissolved

D. bigger

(5) When a river enters its middle and lower course there is a lot of finer material eroded from further upstream which will be carried in _____ .

A. solution

B. suspension

C. saltation

D. traction

(6) Why is the erosion mainly directed vertically and headward within the sediment supply zone?

(7) What kinds of landforms can be developed when corrosion and attrition occur?

(8) What does the way of material transport in a river depend on?

Unit 4

Minerals

Introduction to Minerals

1 A mineral is a naturally occurring, **inorganic**, **crystalline** solid, with a narrowly defined chemical composition and characteristic physical properties. Most minerals are chemical compounds; that is, they consist of two or more elements in combination. Of course there are exceptions, such as gold, copper, **sulphur**, and carbon, which may occur as elements by themselves as well as in chemical compounds[1]. Minerals are naturally occurring substances. This statement rules out[2] laboratory creations. Minerals have a reasonably definite chemical composition. Since they are naturally occurring substances, and not laboratory products, only rarely are they chemically pure compounds. For this reason, such properties as color may vary over a range as wide as from black to white, depending on the percentage of elements present for any mineral[3].

2 Minerals also have certain physical properties, determined by their chemical composition and by the geometric arrangement of the atoms composing them. It is this atomic arrangement that determines the crystal form of a mineral. Other properties include such

things as color, hardness, and specific gravity.

3　The criterion "naturally occurring" excludes from[4] minerals manufactured by humans. Accordingly, most geologists do not regard **synthetic** diamonds and **rubies** and a number of other artificially synthesized substances as minerals.

4　Some geologists think the term "inorganic" in the mineral definition is **superfluous**. It does remind us that animal matter and vegetable matter are not minerals. Nevertheless, some organisms, including **corals**, **clams**, and a number of other animals, construct their shells of the compound calcium carbonate ($CaCO_3$), which is either the mineral **aragonite** or **calcite**.

5　By definition minerals are "crystalline solids", in which the **constituent** atoms are arranged in a regular, three-dimensional framework, as in the mineral **halite**. Under ideal conditions, such as in a **cavity**, mineral crystals can grow and form perfect crystals that possess **planar** surfaces (crystal faces), sharp corners, and straight edges. In other words, the regular geometric shape of a well-formed mineral crystal is the exterior manifestation of an ordered internal atomic arrangement. Not all rigid substances are crystalline solids; for example, natural and manufactured glass lack the ordered arrangement of atoms and are said to be **amorphous**, meaning "without form".

6　Crystalline structure can be demonstrated even in minerals lacking obvious crystals. Many minerals possess a property known as **cleavage,** meaning that they break or split along closely spaced, smooth planes. The fact that these minerals can be split along such smooth planar surfaces indicates that the mineral's internal structure controls such breakage[5].

7　Mineral composition is generally shown by a chemical **formula**, which is a shorthand way of indicating the numbers of atoms of different elements composing a mineral[6]. The definition of a mineral contains the phrase "a narrowly defined chemical composition" because some minerals actually have a range of compositions. For many minerals, the chemical composition is constant: Quartz is always composed of silicon and oxygen (SiO_2), and halite contains only **sodium** and **chlorine** (NaCl). Other minerals have a range of compositions because one element can substitute for[7] another if the atoms of two or more elements are nearly the same size and the same charge.

8　The last criterion in the definition of a mineral – "characteristic physical properties", refers to such properties as hardness, color, and external crystal form, which are controlled by composition and structure.

9   The property of "**scratchability**", or hardness, can be tested fairly reliably. For a true test of hardness, the harder mineral or substance must be able to make a **groove** or scratch on a smooth, fresh surface of the softer mineral. The softest mineral, **talc**, is used for talcum powder because of its softness; while diamond is the hardest natural substance on the earth.

10 Color is so obvious that beginners tend to rely too heavily on it as a key to mineral identification. Unfortunately, color is also apt to[8] be the most ambiguous physical properties. Color is extremely variable in quartz and many other minerals because even **minute** chemical impurities can strongly influence it.

11 The crystal form of a mineral is a set of faces that have a definite geometric relationship to one another. What most people call a "crystal" is a mineral with well-developed crystal faces. If minerals were always able to develop their characteristic crystal forms, mineral identification would be a much simpler task. In rocks, however, most minerals grow while competing for space with other minerals. In fact, the orderly sets of faces that make up a crystal form can develop only under rather specialized conditions. Specifically, most minerals are able to develop their characteristic crystal faces only if they are surrounded by a fluid that can be easily displaced as the crystal grows. On the other hand, a few minerals, notably **garnet**, can overpower and displace surrounding solid material during growth, so that they almost always develop their characteristic crystal faces.

12 In summary, then, a mineral may be defined as a naturally occurring substance with a fairly definite chemical composition and characteristic physical properties by which it may be identified. In short, a typical mineral is a crystalline solid and is an inorganic substance. Most are chemical compounds, but a few, such as the diamond, may consist of a single element.

13 Before we discuss the characteristics of individual minerals, we should learn of the essential properties which are the chief means of the identification. Physical properties are the things we can see, feel, or, for such minerals as halite (rock salt), taste. Truly enough, the chemical composition is possibly the most **diagnostic** property a mineral possesses, but few of us are going to pack along a fully equipped chemical laboratory to be used for mineral identification on a field trip[9]. Since one of the critical differences between minerals and rocks is that minerals are approximately **homogeneous** substances, and most rocks are not, this means that one piece of quartz will be about as hard as another piece,

that it will have the same specific gravity, and if formed in a similar environment, it will have about the same crystal form[10]. Most rock deposits contain metals or minerals, but when the **concentration** of valuable minerals or metals is too low to justify mining, it is considered a waste or **gangue** material (Figure 4-1).

Figure 4-1　Garnet in Metamorphic Rock

New words

inorganic /ˌɪnɔːˈgænɪk/	*adj.* 无机的；无生物的
crystalline /ˈkrɪstəlɑɪn/	*adj.* 结晶的；水晶般的
sulphur /ˈsʌlfə(r)/	*n.* 硫黄；硫黄色
	vt. 使硫化；用硫黄处理
synthetic /sɪnˈθetɪk/	*adj.* 综合的；合成的；人造的
	n. 合成物
ruby /ˈruːbɪ/	*n.* 红宝石
superfluous /suːˈpɜːfluəs/	*adj.* 多余的；不必要的
coral /ˈkɒrəl/	*n.* 珊瑚；珊瑚虫
	adj. 珊瑚的；珊瑚色的
clam /klæm/	*n.* 蛤蜊
aragonite /ˈərægənaɪt/	*n.* [矿物] 霰石
calcite /ˈkælsɑɪt/	*n.* 方解石

constituent /kənˈstɪtjuənt/	*n.* 成分，组成部分
	adj. 构成的；选举的
halite /ˈhælaɪt/	*n.* 岩盐；石盐
cavity /ˈkævətɪ/	*n.* 洞；空穴
planar /ˈpleɪnə/	*adj.* 平面的；平坦的；二维的
amorphous /əˈmɔːfəs/	*adj.* 无定形的；非晶质的
cleavage /ˈkliːvɪdʒ/	*n.* [矿物] 解理；劈理
formula /ˈfɔːmjələ/	*n.* [数] 公式，准则；配方
sodium /ˈsəudɪəm/	*n.* [化] 钠（11 号化学元素，符号为 Na）
chlorine /ˈklɔːriːn/	*n.* [化] 氯（17 号化学元素，符号为 Cl）
scratchability /skrætʃəˈbɪlɪtɪ/	*n.* 柔软度
groove /gruːv/	*n.* 沟；槽
	vt. 开槽于
	vi. 形成沟槽
talc /tælk/	*n.* 滑石；云母
	vt. 用滑石粉处理；在……上撒滑石粉
minute /maɪˈnjuːt/	*adj.* 微小的；详细的；细致的；精密的
garnet /ˈgɑːnɪt/	*n.* [矿物] 石榴石；深红色
	adj. 深红色的；暗红色的
diagnostic /ˌdaɪəgˈnɒstɪk/	*n.* 诊断法；诊断结论
	adj. 诊断的；特征的
homogeneous /ˌhɒməˈdʒiːnɪəs, həu-/	*adj.* 均质的；均匀的
concentration /ˌkɒnsnˈtreɪʃn/	*n.* 浓度；集中；浓缩
gangue /gæŋ/	*n.* [地质] 脉石；矿石；尾矿

Notes to the text

(1) Of course there are exceptions, such as gold, copper, sulphur, and carbon, which may occur as elements by themselves as well as in chemical compounds.

which...是一个非限制性定语从句，对前面列举的矿物做补充说明。

译文：当然，也有例外情况，像金、铜、硫和碳，它们既可作为元素单独产出，也可以化合物的形式出现。

(2) rule out：排除；排斥；不考虑

(3) For this reason, such properties as color may vary over a range as wide as from black to white, depending on the percentage of elements present for any mineral.

over a range (of sth.)：在……范围

depending on...：depending on 是现在分词做 color 的后置定语。

译文：因此，像颜色这样的特性就可能在黑色到白色一个很宽的范围内变化，这取决于某一种矿物中存在的元素的百分比。

(4) exclude from：排除；排斥

(5) The fact that these minerals can be split along such smooth planar surfaces indicates that the mineral's internal structure controls such breakage.

其中 that 引导 the fact 的同位语从句。

译文：这些矿物能沿着这些平滑面分开，说明矿物的内部结构控制了矿物的裂离。

(6) Mineral composition is generally shown by a chemical formula, which is a shorthand way of indicating the numbers of atoms of different elements composing a mineral.

composing a mineral：composing 是现在分词做 different elements 的后置定语。

译文：矿物成分通常用化学式表示，它能快捷地表明构成这种矿物不同元素的原子数量。

(7) substitute for：代替；取代

(8) be apt to：倾向于

(9) Truly enough, the chemical composition is possibly the most diagnostic property a mineral possesses, but few of us are going to pack along a fully equipped chemical laboratory to be used for mineral identification on a field trip.

a mineral possesses 是 the most diagnostic property 的定语从句，省略了关系代词 that。

译文：不错，化学成分也许是矿物所具有的最主要的鉴别特质，但是我们却很少有人打算把全套设备的化学实验室搬到野外去鉴定矿物。

(10) Since one of the critical differences between minerals and rocks is that minerals are approximately homogeneous substances, and most rocks are not, this means that one piece of quartz will be about as hard as another piece, that it will have the same specific gravity, and if formed in a similar environment, it will have about the same crystal form.

此处的 that 与 this means that 中的 that 并列引导 means 的两个宾语从句，即 and if formed 和 and if one piece of quartz is formed。

译文：由于矿物和岩石的一个重大区别在于矿物是近于各向同性的物质，而大部分岩石则不是。因此，这就意味着每块石英的硬度大致相同，相对密度相同，而且如果是在相似的环境中形成，那么它们还会有大致相同的晶体形状。

Understand the text

Answer the following questions according to the passage you have read.

(1) What characteristics does a mineral have?

(2) Are minerals chemically pure compounds? Why or why not?

(3) What are physical properties of minerals determined by?

(4) Do you think the term inorganic in the mineral definition is superfluous? Why or why not?

(5) Does crystalline structure only exist in minerals with crystals?

(6) How is mineral composition generally shown?

(7) What are the physical properties of a mineral controlled by?

(8) Is color reliable to identify a mineral? Why or why not?

(9) How do most minerals develop their crystal faces?

(10) How is a mineral defined?

Translation

1. Translate the following sentences into Chinese.

(1) Since they are naturally occurring substances, and not laboratory products, only rarely are they chemically pure compounds.

_____

_____

_____

_____

(2) Minerals also have certain physical properties, determined by their chemical composition and by the geometric arrangement of the atoms composing them.

_____

_____

_____

_____

_____

(3) In other words, the regular geometric shape of a well-formed mineral crystal is the exterior manifestation of an ordered internal atomic arrangement.

_____

_____

_____

_____

(4) Color is extremely variable in quartz and many other minerals because even minute chemical impurities can strongly influence it.

_____

_____

_____

_____

(5) Most rock deposits contain metals or minerals, but when the concentration of valuable minerals or metals is too low to justify mining, it is considered a waste or gangue material.

_____

_____

_____

_____

_____

2. Translate the following passage into English.

　　由于矿物与岩石的一个重大区别在于矿物是近于各向同性的物质，而大部分岩石则不是，因此这就意味着每块石英的硬度大致相同、相对密度相同，而且如果在相似环境中形成，则其晶体形状也大致相同。

_____
_____
_____
_____
_____
_____
_____
_____
_____
_____

Writing skill

Principles of Writing an Outline

An outline is a direct and clear map of your research paper. It shows what each paragraph contains, in what order paragraphs appear, and how all the points fit together as a whole. Most outlines use bullet points or numbers to arrange information and convey points. Outlining is a vital part of planning process for a research paper . It allows you to brainstorm new ideas and make sure the paper will be organized, focused, and supported. It will be much easier to write from an outline instead of starting from a blank page.

There are few principles that are to be followed to make the outlining process more efficient and effective.

1. Coordination

Coordination is the principle of outlining a topic into groups of thoughts in a balanced manner. Basically, the titles of the same level need to be of the same value. By the same value it means the same significance for the essay text. For example, if the topic to write about is minerals, then chemical compositions and physical properties will be two parallel categories having subcategories and being equal by their significance for the essay.

2. Division

Together with main points having the same significance, there should be more subpoints, which are more concrete and specific. Those points should appear as subcategories for the main ones. You can select any possible way to indicate those categories: either letters in alphabetical order (a, b, c and so on), or numbers with dots (1.1, 1.2, etc.). The only thing to remember is that there should be at least two subcategories to start a division. There cannot be only A without B point following.

3. Subordination

Subordination is the way of relationship between the points being a result of division. Subordination means that the subpoints included in an outline of a research paper need to be equal in significance one for another and have the relationship to their main point. For example:

Features of Minerals

1 Physical Properties

1.1 Atomic Arrangement

1.2 Color

1.3 Hardness

2 Chemical Compositions

2.1 Single Chemical Element

2.2 Chemical Compounds

It might be noticed that there are different levels under the general topic "Features of Minerals". The first level goes more general--from "physical properties" to "chemical compositions". The second level in this outline has several points respectively. The first main point can be broken down further into 3 sub-points: atomic arrangement, color and hardness, while the second one includes 2 sub-points: single chemical element and chemical compounds.

4. Consistency

Categories of the same level need to have similar format. There can be the following types of format of different points: sentences, phrases or fragments. The outline can be organized in whatever format fits into the structure needed for the type of paper being written. One common outline format uses Roman numerals, letters, and numbers. Other outlines can use bullet points or other symbols. It is advisable to use whatever organizational patterns work best for the research paper, as long as the writer understands his own organizational tools. Outlines can be written using complete sentences or fragments. Whatever format to be chosen, the writer needs to be consistent in the process of outlining.

Writing practice

You may try to write a draft outline of a research paper based on Text A under the guidance of the principles directed above.

Text B

Important Minerals

1　It is useful to be able to associate the names of important minerals with the physical properties that identify them. Of course, what constitutes an "important" mineral depends on your perspective. To a miner or prospector, an important mineral is one that is commercially valuable (and is, by implication, relatively uncommon). A "rock hound"[1] is interested in collecting any mineral that is pretty or unusual. A gemologist specializes in those varieties of minerals that are of gem quality (diamonds, emeralds[2], etc.). A mineralogist is a scientist who studies the chemistry and crystallographic structure[3] of minerals. The minerals which are regarded as important are those that help us understand the nature of the earth. Rock-forming minerals are particularly attractive because they make up most of the rocks of the earth's crust.

2　More than 3,500 minerals have been identified and described on the earth, most of which are rare and not important to geology (many occur at only a single site on the globe). Only a few hundred – perhaps two dozen – are particularly common and classified as rock-forming minerals. Considering that 92 naturally occurring elements have been discovered, one might think that an extremely large number of minerals could be formed, but several factors limit the number of possible minerals. For one thing, many combinations of elements are chemically impossible; no compounds are composed of only potassium[4] and sodium or of silicon and iron, for example. Another important factor restricting the number of common minerals is that only eight chemical elements make up the bulk of the earth's crust. Oxygen and silicon constitute more than 74% (by weight) of the crust and nearly 84% of the atoms available to form compounds. By far, the most common minerals in the crust consist of silicon and oxygen, combined with one or more of the other elements.

3　Quartz may be the only familiar name among the most common minerals, unless you have already had some exposure to geology. However, like people, each mineral has its own character or physical properties.

4　Minerals with similar crystal structures and compositions are grouped under a common name. For instance, the feldspar[5] group, the most common and abundant minerals composing over 60% of the rock materials in the crust, all has similar crystal structures of oxygen, silicon, and aluminum atoms. Minerals within the group are named according to whether potassium (orthoclase[6] or potassium feldspar[7]) or sodium and calcium (plagioclase feldspar[8]) are incorporated into[9] this basic crystal structure.

5　The pyroxene[10] group and the amphibole[11] group, which are single- and double-chain silicates[12], respectively, contain a number of minerals. Augite[13] is the most common pyroxene, and hornblende[14] is the most common amphibole.

6　The mica group is characterized by minerals with a sheet silicate structure. This distinctive property results from a single, perfect, cleavage plane, which is repeated throughout the mineral. The two most common micas are biotite[15] and muscovite[16]. Biotite is a dark-colored, iron/magnesium[17]-bearing mica. Muscovite mica lacks iron and magnesium and is transparent or white.

7　Calcite is the only common rock-forming mineral which lacks silicon. All others are silicon and oxygen compounds in the form of silica or silicates. Calcite, however, is a carbonate, a class of compounds based on carbon-oxygen combinations. At first glance, calcite might be confused with quartz because both are clear, colorless and glassy. However, unlike quartz, calcite is quite soft, has excellent cleavage and fizzes[18] when drops of hydrochloric acid[19] touch it. Calcite is a very wide-spread mineral in certain common rocks and in the shells of many organisms.

Notes to the text

(1) rock hound：石油勘探地质学家

(2) emerald /ˈemərəld/ *n.* 祖母绿；绿宝石；翡翠

(3) crystallographic structure：晶体结构

(4) potassium /pəˈtæsiəm/ *n.* 钾

(5) feldspar /ˈfeldspɑː/ *n.* 长石

(6) orthoclase /ˈɔːθəʊkleɪz/ *n.* 正长石

(7) potassium feldspar：钾长石

(8) plagioclase feldspar：斜长石

(9) incorporate into：并入

(10) pyroxene /paɪəˈrɒksiːn/ *n.* 辉石；锂辉石

(11) amphibole /ˈæmfɪbəʊl/ *n.* 闪石；角闪石

(12) double-chain silicate：双链型硅酸盐

(13) augite /ˈɔːdʒaɪt/ *n.* 普通辉石；斜辉石

(14) hornblende /ˈhɔːnblend/ *n.*（普通）角闪石

(15) biotite /ˈbaɪətaɪt/ *n.* 黑云母

(16) muscovite/ˈmʌskəʊvaɪt/ *n.* 白云母

(17) magnesium /mægˈniːziəm/ *n.* 镁

(18) fizz /fiz/ *n.* 发嘶嘶声；起泡沫

(19) hydrochloric acid：盐酸

Questions for review

(1) The term "crystalline" in the definition of a mineral means that a mineral _____.

    A. has an atomic number of at least 92

    B. possesses more protons than neutrons

    C. is characterized by physical properties such as hardness and cleavage

    D. has an orderly internal arrangement of atoms

(2) Minerals that break along smooth internal planes of weakness have the property known as _____.

    A. cleavage

    B. double refraction

    C. specific gravity

    D. streak

(3) The two common rock-forming minerals from the carbonate group are _____.

    A. hematite and magnetite

    B. calcite and dolomite

    C. clay and feldspar

    D. halite and gypsum

(4) The two most abundant elements in the earth's crust are _____.

    A. iron and calcium

    B. oxygen and silicon

    C. potassium and magnesium

    D. hydrogen and nitrogen

(5) What factors restrict the formation of possible minerals according to the text?

(6) How do you group minerals?

(7) How many groups of common minerals are mentioned in this text? What are they?

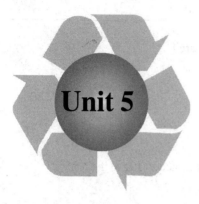

Unit 5

Crystalline Rocks

Igneous Rocks

1 **Igneous** rocks, one of the three main rock types, the other being **metamorphic** and **sedimentary** rocks, are produced by cooling and **solidification** at molten rock-making material called **magma** or lava[1]. Magma is molten rock that is formed in very hot conditions inside the earth with or without suspended crystal and gas bubbles; while lava is the very hot liquid rock that comes out of a volcano to reach the surface (magma is called lava when it reaches the earth's surface). Igneous rocks may form with or without crystallization, either below the surface as **intrusive** (**plutonic**) rocks or on the surface as **extrusive** (volcanic) rocks (Figure 5-1). This magma can be derived from partial melts of pre-existing rocks in either a planet's mantle or crust. Typically, the melting is caused by one or more of three processes: an increase in temperature, a decrease in pressure, or a change in composition. Dissolved gases are important constituents of magma and lava but tend to be excluded from the rock-making minerals when the hardening process occurs. Hardening may take place within or upon the earth's crust, which may result in the subdivision of intrusive and extrusive groups. **Granite**, **basalt** and **obsidian** are familiar examples among over 700 types of igneous rocks described, most of which have formed beneath the surface of the earth's crust.

Figure 5-1   Extrusive (Volcanic) Rocks

2 Igneous rocks are generally classified first on the basis of their formation (and therefore their texture) as intrusive or extrusive.

3 Intrusive rocks are formed from magma that cools and **solidifies** within the crust of a planet. Surrounded by pre-existing rock (called country rock[2]), the magma cools slowly; as a result, these rocks are coarse-grained. Intrusive rocks can also be classified according to the shape and size of the intrusive body and its relation to the other formations into which it intrudes. Intrusive rock masses are younger than the rocks they intrude, and they are exposed in places at the surface today because erosion has removed the older rocks which once were around and on top of them.

4 Bodies of intrusive rocks are exposed only after erosion and, usually, uplift. The central cores of major mountain ranges consist of intrusive igneous rocks, usually granite. When exposed by erosion, these cores may occupy huge areas of the earth's surface. Coarse-grained intrusive igneous rocks which form at depth within the crust are termed as **abyssal** rocks[3] and the ones which form near the surface are termed as **hypabyssal** rocks[4]. Granite is a coarse-grained rock composed predominantly of feldspar and quartz, which is the chemical and mineralogical equivalent of **rhyolite** and the most abundant intrusive rock found in the continents.

5 Some intrusive rocks are fine-grained, like their extrusive counterparts. These are formed from magma that cools rapidly. Apparently, they intrude relatively cool country rock near

the surface of the earth (at depths of probable less than 2 kilometers). If a rock is fine-grained, even if it solidifies near the surface, it has the same name as the corresponding extrusive rock. For example, a fine-grained rock of intermediate composition is an **andesite** whether it is formed from a lava flow or from rapid cooling near the surface. Igneous rocks that are formed at considerable depth–usually more than several kilometers–are called plutonic rocks. Characteristically, these rocks are coarse-grained, reflecting the slow cooling and solidification of magma.

6　Extrusive igneous rocks are formed at the crust's surface as a result of the partial melting of rocks within the mantle and crust. They cool and solidify quicker than intrusive igneous rocks and the crystals that form do not have time to grow very large, thus most extrusive rocks are finely grained. If the cooling has been so rapid as to prevent the formation of even small crystals after extrusion, the resulting rock may be mostly glass (such as the rock obsidian). If the cooling of the lava happened slowly, the rocks would be coarse-grained. The term includes fine-grained crystalline or glassy rocks formed from hot lava **quenched** at or near the earth's surface, and those made of **welded** fragments of ash and glass ejected into the air during a volcanic eruption. Because the minerals are mostly fine-grained, it is much more difficult to distinguish between different types of extrusive rocks than between different types of intrusive rocks. Generally, the mineral constituents of fine-grained extrusive igneous rocks can only be determined by examination of thin section[5] of the rock under a microscope, which means that only an approximate classification can usually be made in the field.

7　Igneous rocks can additionally be classified according to mode of occurrence[6], texture, mineralogy, chemical composition and the geometry of the igneous body. Two important variables used for the classification of igneous rocks are particle size, which largely depends upon the cooling history, and the mineral composition of the rock. The completed rock analysis is first to be interpreted in terms of the rock-forming minerals which might be expected to be formed when the magma crystallizes, e.g., feldspars, quartz or **feldspathoids, olivines**, pyroxenes, amphiboles, micas and so on, which are all important minerals in the formation of almost all igneous rocks[7]. The rocks are divided into groups strictly according to the relative proportion of these minerals to one another. Types of igneous rocks with other essential minerals are very rare, and these rare rocks include those with essential carbonates.

8　In the simplified classification, igneous rock types are separated on the basis of the type of feldspar present, the presence or absence of quartz, and in rocks with no feldspar or quartz, the type of iron or magnesium minerals present. For example, rocks containing quartz (silica in composition) are silica-**oversaturated** and rocks with feldspathoids are silica-**undersaturated**, because feldspathoids cannot coexist in a stable association with quartz. Igneous rocks which have crystals large enough to be seen by the naked eye are called **phaneritic**; those with crystals too small to be seen are called **aphanitic**. Generally speaking, phaneritic implies an intrusive origin while aphanitic an extrusive one.

New words

igneous /ˈɪgnɪəs/	adj. 火的；［岩］火成的；似火的
metamorphic /metəˈmɔːfɪk/	adj. 变质的；变形的
sedimentary /ˌsedɪˈmentri/	adj. 沉积的；沉淀的
solidification /səˌlɪdɪfɪˈkeɪʃn/	n. 凝固，固化；浓缩
magma /ˈmægmə/	n. ［地质］岩浆；糊剂
intrusive /ɪnˈtruːsɪv/	adj. 侵入的
plutonic /pluːˈtɒnɪk/	adj. 火成岩的；深成岩的
extrusive /ɪkˈstruːsɪv/	adj. 突出的；喷出的，挤出的
granite /ˈgrænɪt/	n. 花岗岩；花岗石
basalt /ˈbæsɔːlt/	n. ［岩］玄武岩；黑陶器
obsidian /əbˈsɪdɪən, əb-/	n. 黑曜石
solidify /səˈlɪdɪfaɪ/	vt. 固化；团结；凝固
	vi. 固化；凝固
abyssal /əˈbɪsəl/	adj. 深渊的，深海的；深沉的
hypabyssal /ˌhɪpəˈbɪsəl/	adj. 半深成的，浅成的
rhyolite /ˈraɪəlaɪt/	n. ［岩］流纹岩；表面光滑的火山岩
andesite /ˈændɪzaɪt/	n. 安山岩；安山石
quench /kwentʃ/	vt. 熄灭，［机］淬火；冷却
	vi. 熄灭；平息
weld /weld/	n. 焊接

	vt. 焊接；使结合；使成整体
	vi. 焊牢
feldspathoid /ˈfeldspæθɔɪd/	*n.* 似长石
	adj. 似长石的
olivine /ˈɒlɪˈviːn/	*n.* [矿物] 橄榄石；橄榄绿
oversaturated /ˈəuvəˈsætʃəˌreɪtɪd/	*adj.* [化] 过饱和的；硅石的
undersaturated /ˌʌndəˈsætʃəreɪtɪd/	*adj.* 未饱和的；不饱和的
phaneritic /ˌfænəˈrɪtɪk/	*adj.* 粗晶的；显晶岩的
aphanitic /ˌæfəˈnɪtɪk/	*adj.* 非显晶岩的；隐晶岩的

Notes to the text

(1) Igneous rocks, one of the three main rock types, the other being metamorphic and sedimentary rocks, are produced by cooling and solidification at molten rock-making material called magma or lava.

本句中，one of the three main rock types 做 igneous rocks 的同位语；the other being metamorphic and sedimentary rocks 是独立主格结构，补充说明其他两种主要岩石；called magma or lava 做 material 的后置定语。

译文：作为三大主要岩石之一的火成岩是由称为岩浆或熔岩的熔融状造岩物质经过冷却和凝固而成的，其他两种（主要岩石）是变质岩和沉积岩。

(2) country rock：围岩；母岩

(3) abyssal rocks：深成岩

(4) hypabyssal rocks：浅成岩；半深成岩

(5) thin section：薄片；薄剖面

(6) mode of occurrence：产状；赋存状态

(7) The completed rock analysis is first to be interpreted in terms of the rock-forming minerals which might be expected to be formed when the magma crystallizes, e.g., feldspars, quartz or feldspathoids, olivines, pyroxenes, amphiboles, micas and so on, which are all important minerals in the formation of almost all igneous rocks.

译文：完整的岩石分析首先要用造岩矿物来解释，在这些造岩矿物，如长石、石英、似长石、橄榄石、辉石、闪石和云母等主要形成于岩浆结晶时，它们几乎是形成所有火成岩的重要矿物。

Understand the text

Answer the following questions according to the passage you have read.

(1) How are igneous rocks formed?

(2) What is the difference between magma and lava?

(3) What are the characteristics of intrusive rocks and extrusive rocks?

(4) What are the differences between intrusive rocks and extrusive rocks?

(5) Are granite and rhyolite the same? Why or why not?

(6) What is the rock formed at considerable depth called?

(7) What factors can be based on to classify igneous rocks?

(8) Why are rocks with feldspathoids silica-undersaturated?

(9) What kind of igneous rocks are called phaneritic?

(10) What are the differences between phaneritic and aphanitic?

Translation

1. Translate the following sentences into Chinese.

(1) This magma can be derived from partial melts of pre-existing rocks in either a planet's mantle or crust.

_____

_____

_____

_____

_____

(2) Dissolved gases are important constituents of magma and lava but tend to be excluded from the rock-making minerals when the hardening process occurs.

_____

_____

_____

_____

(3) Granite is a coarse-grained rock composed predominantly of feldspar and quartz, which is the chemical and mineralogical equivalent of rhyolite and the most abundant

intrusive rock found in the continents.

_____

_____

_____

(4) The extrusive rock includes fine-grained crystalline or glassy rocks formed from hot lava quenched at or near the earth's surface, and those made of welded fragments of ash and glass ejected into the air during a volcanic eruption.

_____

_____

_____

_____

(5) In the simplified classification, igneous rock types are separated on the basis of the type of feldspar present, the presence or absence of quartz, and in rocks with no feldspar or quartz, the type of iron or magnesium minerals present.

_____

_____

_____

_____

2. Translate the following passage into English.

火成岩也称岩浆岩，来源于拉丁文 ignis，意为火焰，一般是由地下深处炽热的熔融体和熔岩在地下或地表冷却凝固而成的。火成岩分为侵入岩和喷出岩。现在已经发现 700 多种类型的岩浆岩，它们中的大部分形成于地壳表面以下。花岗岩、安山岩及玄武岩是常见的例子。一般来说，火成岩易沿一些大陆板块边缘的火山区生成。

_____

_____

_____

_____

_____

_____

_____

_____

_____

_____

_____

_____

Writing skill

Steps of Writing an Outline

When you begin to write an outline for a research paper，there are two general formats of outline you can take: the topic outline or the sentence outline.

The **topic outline** is composed of words and short phrases. This kind of outline is useful when you are dealing with a number of different issues that could be arranged in a variety of different ways in your paper. Because of short phrases having brief and clear view, writing the topic outline will be easier and faster.

The **sentence outline** is conducted in full sentences. This type of outline is useful when your paper focuses on complex issues in detail. The sentence outline is also useful because sentences themselves have many of the details and it allows you to include those details in your paper instead of spending much time composing sentences from short phrases.

An outline is a framework for presenting the main and supporting ideas for a particular topic, organizing these ideas to build the argument towards an evidence-based conclusion. A strong outline will also help you to develop a logical, coherent structure for your paper.

Here are five steps to a strong outline:

1. Identify the research problem and the purpose of your paper.

The research problem is the focal point from which the rest of the outline flows. Having an objective in mind will help you set guidelines and limitations on what is appropriate content for your essay.

2. Brainstorm a list of main ideas.

Review all your note cards and decide what points you would like to discuss in your paper. The goal is to come up with a list of essential ideas that you are planning to present in your article or essay.

3. Organize those main ideas.

In order to make sense to you and your reader, the main ideas you collect can be arranged in different logical, numerical modes, such as division and classification; comparison and contrast; cause and effect, etc..

4. Create supporting ideas.

Provide support for the main point. The number of supporting ideas that you use relies on the amount of information that you attempt to cover.

5. Form a logical preliminary outline.

Review and revise the outline to make sure that you've included all of your ideas and each point connects back to your main point.

Here is a sample outline of Text A:

Igneous Rock

1 Introduction to Igneous Rock

1.1 Formation of Igneous Rock

1.2 Places of Formation

1.2.1 Below the Surface

1.2.2 On the Surface

1.3 Melting Process

1.4 Hardening Process

2 Two Major Igneous Rocks

2.1 Intrusive Rocks

2.1.1 Definition of Intrusive Rocks

2.1.2 Characteristics of Intrusive Rocks

2.2 Extrusive Rocks

2.2.1 Definition of Extrusive Rocks

2.2.2 Characteristics of Extrusive Rocks

Writing practice

You may try to write an outline based on the research topic you are interested in.

Text B

Metamorphic Rocks

1　Metamorphic rocks arise from the transformation of existing rock types in a process called metamorphism[1] (a word from Latin and Greek which means literally "changing of form"), referring to changes to rocks that take place in the earth's interior. The changes may be new textures, new mineral assemblages or both with transformation occurring in the solid state (meaning the rock does not melt).

2　The new rock, the metamorphic rock, in nearly all cases has a texture clearly different from that of the original rock, or parent rock[2]. The metamorphic rocks come into being through the transformation of protolith[3] such as sedimentary, igneous or older metamorphic rocks which are subjected to heat (temperature greater than 150℃ to 200℃) and pressure into new and notably different types. New kinds of minerals or mineral particles with different shapes or orientations can be produced in the crust by heat, pressure and the chemical action of solutions. When limestone is metamorphosed[4] to marble, for example, the fine grains of calcite coalesce[5] and recrystallize into larger calcite crystals. The calcite crystals are interlocked[6] in a mosaic pattern[7] that gives marble a texture distinctly different from that of the parent limestone. If the limestone is composed entirely of calcite (without impurities), metamorphism into marble involves no new minerals, only a change in texture.

3　More commonly, the various elements of a parent rock react chemically and crystallize into new minerals, thus making the metamorphic rock distinct both mineralogically and texturally from the parent rock. This is because the parent rock is unstable in its new environment. For example, clay minerals form at the earth's surface; therefore, they are stable at low temperature and pressure conditions both at and just below the earth's surface. When subjected to the temperatures and pressures deep within the earth's crust, the clay minerals of shale can recrystallize into coarse-grained mica.

4　Metamorphic rocks make up a large part of the earth's crust and are classified by texture and by chemical and mineral assemblage[8] (metamorphic facies[9]). They may be formed simply by being deep beneath the earth's surface, subjected to high temperatures and great pressure of the rock layers above it. They can form from tectonic processes[10] such as continental collisions[11], which cause horizontal pressure, friction[12] and distortion. They are also formed when rock is heated up by the intrusion of hot molten rock called magma from the earth's interior. Some examples of metamorphic rocks are gneiss[13], slate[14], marble, schist[15] and quartzite[16] (Figure 5-2) .

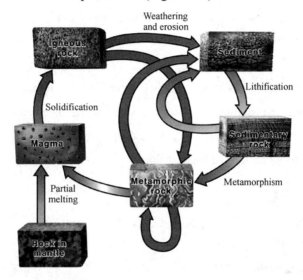

Figure 5-2   The Forming Process of Metamorphic Rocks

5 A metamorphic rock owes its characteristic texture and particular mineral content to several factors, the most important being (1) the composition of the parent rock before metamorphism; (2) temperature and pressure during metamorphism; (3) the effects of tectonic forces; and (4) the effects of fluids, such as water. Usually no new elements or chemical compounds are added to the rock during metamorphism except perhaps water; therefore, the mineral content of the metamorphic rock is controlled by the chemical composition of the parent rock. For example, a basalt always metamorphoses into a rock in which the new minerals can collectively accommodate[17] the approximately 50% silica and relatively high amounts of the oxides[18] of iron, magnesium, calcium and aluminum in the original rock. On the other hand, a limestone, composed essentially of calcite, cannot metamorphose into a silica-rich rock.

6 A mineral is said to be stable if, given enough time, it does not react with or convert to a

new mineral or substance. Any mineral is stable only within a given temperature range. The stable temperature range of a mineral varies with factors such as pressure and the presence or absence of other substances. Minerals stable at higher temperatures tend to be less dense (or have a lower specific gravity) than chemically identical minerals stable at lower temperatures. As temperature increases, the ions[19] vibrate more within their sites in the crystal structure. A more open (less tightly packed) crystal structure, such as high-temperature minerals tend to have, allows greater vibration of ions. If the heat and resulting vibrations become too great, the crystal breaks apart and the substance becomes liquid.

7　The kind of metamorphic rock that forms is determined by the metamorphic environment, primarily the particular combination of pressure, stress and temperature, and by the chemical constituents of the parent rock. Many kinds of metamorphic rocks exist because of many possible combinations of these factors.

Notes to the text

(1) metamorphism /metəˈmɔːfɪz(ə)m/ *n.* 变质；变性

(2) parent rock：母岩，原生岩

(3) protolith /ˌprəutəulɪθ/ *n.* 原岩

(4) metamorphose /ˌmetəˈmɔːfəuz/ *n.* 变质；变形

(5) coalesce /ˌkəuəˈles/ *v.* 使联合；合并

(6) interlock /ˈɪntəlɒk/ *v.* 互锁；连锁

(7) mosaic pattern：镶嵌图案

(8) mineral assemblage：矿物组合

(9) metamorphic facies：变质相

(10) tectonic processes：构造变化过程

(11) continental collisions：大陆碰撞

(12) friction /ˈfrɪkʃn/ *n.* 摩擦；摩擦力

(13) gneiss /naɪs/ *n.* 片麻岩

(14) slate /sleɪt/ *n.* 板岩，高灰煤

(15) schist /ʃɪst/ *n.* 片岩

(16) quartzite /ˈkwɒːtsaɪt/ *n.* 石英岩；硅岩

(17) accommodate /əˈkɒmədeɪt/ *v.* 容纳；适应

(18) oxide /ˈɒksaɪd/ *n.* 氧化物

(19) ion /ˈaɪən/ *n.* 离子

Questions for review

(1) Limestone recrystallizes during metamorphism into _____.

    A. hornfels

    B. marble

    C. quartzite

    D. schist

(2) What are the major factors contributing to metamorphism?

(3) What is metamorphism characterized by?

(4) In what aspects is metamorphic rock distinct from the parent rock?

(5) How may metamorphic rocks be formed?

(6) What does a metamorphic rock mostly owe its characteristic texture and particular mineral content to?

(7) What factors will affect the stability temperature range of a mineral according to the text?

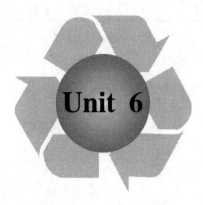

Unit 6

Sedimentation

Sediment

1   Most sedimentary rocks form from loose grains of sediment. Sediment includes such particles as sand on beaches, mud on a lake bottom, boulders frozen into glaciers, pebbles in streams, and dust particles settling out of the air. An accumulation of clam shells on the sea bottom offshore is sediment, as are coral fragments broken from a **reef** by large storm waves[1].

2   Sediment is the collective name for loose, solid particles that originate from: (1) weathering and erosion of preexisting rocks; (2) chemical **precipitation** from solution, including **secretion** by organisms in water. These particles usually gather in layers on the earth's surface. An important part of the definition is that the particles are loose. Sediments are said to be **unconsolidated**, which means that the grains are separate, or unattached to one another.

3   Weathering is an important part of the rock cycle because it yields particles and dissolves substances, both of which might be raw materials for sedimentary rocks, the second major family of rocks. All sedimentary rocks are composed of sediment – solid particles derived by mechanical and chemical weathering, minerals **precipitated** from solution by chemical

processes, or minerals **secreted** by organisms when they build their skeletons. Thus all sediments are derived by one weathering process or another, eroded from the weathering site, transported elsewhere, and deposited as a loose **aggregate** of particles. Sand and gravel in stream channels and on beaches as well as mud on the seafloor are examples of sedimentary deposits.

4 Loose sediments may be changed into solid sedimentary rocks by pressure from overlying strata which are deposited later and by the precipitation of cementing material as a **binder** around individual grains[2]. Most sedimentary rocks form when sediment is transformed into solid rock, but a few skip the unconsolidated sediment stage and form directly as solids. The presence of different layers, beds, or **strata** constitutes an outstanding feature of most sedimentary rocks and serves to distinguish them from many igneous and metamorphic rocks.

5 One important criterion for classifying sedimentary particles is their size, particularly for solid particles, or **detrital** sediment, derived by weathering as opposed to chemical sediment, and minerals extracted from solution by inorganic chemical processes, or used in the skeletons of organisms[3]. Particles described as gravel measure more than 2 mm, whereas sand measures 1/16–2 mm. Composition is not a consideration in naming these particles. Most gravel is made up of rock fragments, that is, small pieces of granite, basalt or any other rock type, and quartz, whereas the most common mineral in sand is quartz, but a number of others might be present as well. Particles smaller than 1/256 mm are termed clay, but clay has two meanings. One is simply a size designation, but it also refers to certain types of sheet silicates known as clay minerals. However, most clay minerals are also clay sized.

6 Weathering and erosion are fundamental processes in the origin of sediment and sedimentary rocks, but so is sediment transport, that is, the movement of detrital and chemical sediment by natural processes. Glaciers can carry sedimentary particles of any size, whereas wind transports only sand and smaller sediment. Waves and marine currents transport sediment along shorelines, but by far the most common way to transport sediment from its weathering site to other locations is by running water.

7 Rounding is the grinding away of sharp edges and corners of rock fragments during transportation. Rounding occurs in sand and gravel as rivers, glaciers, or waves cause

particles to hit and scrape against one another or against a rock surface, such as a rocky stream bed. Boulders in a stream may show substantial rounding in less than one mile of travel. Because rounding during transportation is so rapid, it is a much more important process than **spheroidal** weathering which also tends to round sharp edges.

8　Sorting is a process in which sediment grains are selected and separated according to grain size (or grain shape or specific gravity). Because of its high **viscosity** and manner of flow, a glacier does not sort the sediment it carries. Glaciers deposit all sediment sizes in the same place, so glacial sediment is usually an unsorted mixture of clay, **silt**, sand, and gravel.

9　Regardless of how sediment is transported, it may be transported a considerable distance from its source, and it is eventually deposited. Sediment is deposited when running water, glacial ice, waves, or wind loses energy and can no longer transport its load. Deposition also refers to the accumulation of chemical or organic sediment, such as clam shells on the sea floor, or plant material on the floor of a **swamp**. Such sediments may form as organisms die and their remains accumulate, perhaps with no transportation at all. Small particles of clay and silt might be transported to a lake where the lack of currents allows them to settle, thus forming a layer of mud; likewise, mud might settle on a stream's flood-plain. Larger particles of sand and gravel commonly accumulate in stream channels and on beaches, whereas chemical sediment might be deposited in shallow seawater, where organisms use dissolved substances to make their skeletons. Any of these geographic areas in which sediment is deposited is a depositional environment where physical, chemical, and biological processes **impart** various characteristics to sedimentary deposits.

10　Layers result from changes in conditions as deposition takes place; e.g., velocities of the transporting medium may increase or decrease so that larger or smaller particles are dropped at a given spot. Fossils are abundant in some sedimentary rocks, readily distinguishing them from most igneous and metamorphic specimens; but not all sedimentary rocks contain fossils. Fossils "label" the rocks in which they occur and yield much information about the past.

11　**Conglomerate**, sandstone and shale—formerly gravel, sand and mud or clay, respectively – are common types of sedimentary rocks, and the size of the average particle in the rock is the only criterion used in classification here. On the other hand, limestones are also common and are classified on their mineral content, which is chiefly calcium carbonate in the form of calcite. The calcite may have formed by precipitation from solution or by the accumulation of shell fragments.

New words

reef /ri:f/	*n.* 暗礁，礁石；［地质］矿脉；收帆
	vt. 收帆；缩帆
	vi. 缩帆；收帆
precipitation /prɪˌsɪpɪˈteɪʃn/	*n.* ［化］沉淀（物）；冰雹；坠落
secretion /sɪˈkri:ʃn/	*n.* 分泌；分泌物；分凝
unconsolidated /ˈʌnkənˈsɒlɪdeɪtɪd/	*adj.* 疏松的；松散的；未固结的
precipitate /prɪˈsɪpɪteɪt/	*n.* ［化］沉淀物
	vt. 使沉淀；加速，促成
	vi. ［化］沉淀；冷凝成为雨或雪等
secrete /sɪˈkri:t/	*vt.* 隐藏，藏匿；私吞；［生］分泌
aggregate /ˈæɡrɪɡeɪt/	*n.* 集合体，集料
	v. 集合；聚集；合计
binder /ˈbaɪndə(r)/	*n.* 黏合剂；活页夹；煤层杂质条纹
strata /ˈstrɑ:tə/	*n.* （stratum 的复数）层；地层；阶层
detrital /dɪˈtraɪtəl/	*adj.* 碎屑的；由岩屑形成的
spheroidal /sfɪəˈrɔɪdəl/	*adj.* 类似球体的，球状的
viscosity /vɪˈskɒsɪtɪ:/	*n.* ［物］黏性，［物］黏度
silt /sɪlt/	*n.* 淤泥，粉砂
	vt. 使淤塞；充塞
	vi. 淤塞，充塞；为淤泥堵塞
swamp /swɒmp/	*n.* 沼泽；湿地
	vt. 使陷于沼泽；使沉没；使陷入困境
	vi. 下沉；陷入沼泽；陷入困境
impart /ɪmˈpɑ:t/	*vt.* 给予，传授；告知，透露
conglomerate /kənˈɡlɒmərət/	*n.* ［岩］砾岩，聚合物
	vt. 使聚结；凝聚成团
	vi. 凝聚成团

Notes to the text

(1) An accumulation of clam shells on the sea bottom offshore is sediment, as are coral fragments broken from a reef by large storm waves.

在此引导一个部分倒装句，相当于 and so…。

译文： 近海海底堆积的蛤蜊壳就是沉积物，正如从珊瑚礁上由巨浪打破的珊瑚碎片的堆积一样。

(2) Loose sediments may be changed into solid sedimentary rocks by pressure from overlying strata which are deposited later and by the precipitation of cementing material as a binder around individual grains.

by pressure 和 by the precipitation 是两个并列介词短语做状语。as a binder 是介词短语进一步说明 cementing material。

译文： 在后期沉积下来的上覆沉积物的压实和颗粒间胶结物的结晶沉淀共同作用下，松散的沉积物可以固结转变成沉积岩。

(3) One important criterion for classifying sedimentary particles is their size, particularly for solid particles, or detrital sediment, derived by weathering as opposed to chemical sediment, and minerals extracted from solution by inorganic chemical processes, or used in the skeletons of organisms.

solid particles、detrital sediment 和 minerals 是并列关系。derived by 做 detrital sediment 的后置定语，extracted from 和 used in 一起做 minerals 的后置定语。

译文： 对于沉积颗粒划分的一个重要标准就是颗粒的大小，尤其是固态颗粒或来自物理风化而不是化学风化的碎屑颗粒，以及由无机化学作用形成的矿物颗粒或是构成有机物骨骼的生物碎屑。

Understand the text

Answer the following questions according to the passage you have read.

(1) What is sediment?

(2) Where do those loose, solid particles consisting of sediment originate from?

(3) Why is weathering an important part of the rock cycle?

(4) What materials mentioned in the passage are grouped into sediment?

(5) What are fundamental processes in the origin of sediment and sedimentary rocks?

(6) How many ways of sediment transport are mentioned in the passage and what are they?

(7) Why is glacial sediment usually unsorted?

(8) What does deposition refer to in this text?

Translation

1. Translate the following sentences into Chinese.

(1) All sedimentary rocks are composed of sediment – solid particles derived by mechanical and chemical weathering, minerals precipitated from solution by chemical processes, or minerals secreted by organisms when they build their skeletons.

_____

_____

_____

_____

_____

(2) One is simply a size designation, but it also refers to certain types of sheet silicates known as clay minerals. However, most clay minerals are also clay sized.

_____

_____

_____

_____

_____

(3) Rounding is the grinding away of sharp edges and corners of rock fragments during transportation.

_____

_____

_____

_____

_____

(4) Small particles of clay and silt might be transported to a lake where the lack of currents allows them to settle, thus forming a layer of mud; likewise, mud might settle on a stream's flood-plain.

_____

_____

_____

_____

(5) Fossils "label" the rocks in which they occur and yield much information about the past.

_____

_____

_____

_____

_____

2. Translate the following passage into English.

　　沉积物是通过像流水、海浪、洋流、风和冰川这样一些地质营力进行搬运和沉降或沉淀下来的。有的岩石碎屑呈磨圆状，有的则呈棱角状。这种特征取决于搬运距离的远近和其他因素。

_____

_____

_____

_____

_____

_____

_____

_____

Writing skill

How to Write an Abstract?

An abstract is a condensed statement of the contents of a paper. As a short, concise and highly generalized text, an abstract is viewed as a mini-version or a miniature of the document, summarizing the content of the main body.

A powerful abstract enables readers to identify the basic content of a document quickly and accurately, to determine its relevance to their interests, and thus to decide whether they want to continue reading the entire document or not.

An abstract is the indispensable part of an academic research paper. Although an abstract appears as the first part of your paper, it should be written after you have completed your full paper so that all the basic information should be covered. Generally, a typical abstract will provide the following information:

1. **Background of the research** - briefly refers to the relevant research background to provide a context of your study.

2. **Research objectives** - clearly define the reasons and purposes for conducting this research.

3. **Methods** - precisely state the valid and reliable methods applied to reach your result, or the theoretical or subject scope of the paper.

4. **Findings or results** - logically summarize the main research result and the data you have learned, uncovered or created.

5. **Conclusions** - appropriately illustrate the implications of your findings, or whether more research is needed or any recommendation could be applied in future practice.

Sample Abstract

The Grenville Province on the eastern margin of Laurentia is a remnant of a Mesoproterozoic orogenic plateau that comprised the core of the ancient supercontinent Rodinia. As a protracted Himalayan-style orogen, its orogenic history is vital to understanding Mesoproterozoic tectonics and paleoenvironmental evolution. In this study, we compared two geochemical proxies for crustal thickness: whole-rock $[La/Yb]_N$ ratios of intermediate-to-felsic rocks and europium anomalies (Eu/Eu*) in detrital zircons. We compiled whole-rock geochemical data from 124 plutons in the Laurentian Grenville Province and collected trace-element and geochronological data from detrital zircons from the Ottawa and St. Lawrence River (Canada) watersheds. Both proxies showed several episodes of crustal thickening and thinning during Grenvillian orogenesis. The thickest crust developed in the Ottawan phase (~60 km at ca. 1080 Ma and ca. 1045 Ma), when the collision culminated, but it was still up to 20 km thinner than modern Tibet. We speculate that a hot crust and several episodes of crustal thinning prevented the Grenville hinterland from forming a high Tibet-like plateau, possibly due to enhanced asthenosphere-lithosphere interactions in response to a warm mantle beneath a long-lived supercontinent, Nuna-Rodinia.

Background of the research

Research objective

Methods

Findings

Conclusion

Writing practice

You may read some abstracts of research papers and analyze the components of them.

Text B

Sedimentary Rocks

1　Sedimentary rocks constitute only a small part of the earth's crust, but along with sediments they cover most of the seafloor and about 75% of the continents. Accordingly, sedimentary rocks and sediments are the most commonly encountered the earth materials. Sedimentary rock is rock that has formed from (1) lithification[1] of sediment, (2) precipitation from solution, or (3) consolidation of the remains of plants or animals. These different types of sedimentary rocks are called, respectively, clastic, chemical and organic rocks.

2　Most sedimentary rocks are clastic sedimentary rocks, formed from cemented sediment grains that are fragments of preexisting rocks. The rock fragments can be either identifiable pieces of rock, such as pebbles of granite or shale, or individual mineral grains, such as sand-sized quartz and feldspar crystals loosened from rocks by weathering and erosion. Clay minerals formed by chemical weathering are also considered fragments of preexisting rocks. In most cases, the sediment has been eroded and transported before being deposited. During transportation the grains may have been rounded and sorted.

3　Sedimentary breccia is a coarse-grained sedimentary rock formed by the cementation of coarse, angular fragments of rubble. Because grains are rounded so rapidly during transport, it is unlikely that the angular fragments within breccia have moved very far from their source. Sedimentary breccia might form, for example, from fragments that have accumulated at the base of a steep slope of rock that is being mechanically weathered. Landslide deposits also might lithify[2] into sedimentary breccia.

4　Conglomerate is a coarse-grained sedimentary rock formed by the cementation[3] of rounded gravel. It can be distinguished from breccia by the definite roundness of its particles. Because conglomerates are coarse-grained, the particles may not have traveled far; some transport, however, was necessary to round the particles. Angular fragments that fall from a cliff and then are carried a few miles by a river or by waves are quickly rounded. Gravel that is transported down steep submarine canyons, or carried by glacial ice as till[4], however, can be transported tens or even hundreds of miles before deposition.

5　Sandstone is a medium-grained sedimentary rock formed by the cementation of sand grains. Any deposit of sand can lithify to sandstone. Rivers deposit sand in their channels and wind piles up sand into dunes. Waves deposit sand on beaches and in shallow water. Deep-sea currents spread sand over the sea floor. Sandstones show a great deal of variation in mineral composition, degree of sorting and degree of rounding. Sandstones may contain a substantial amount of matrix[5], fine-grained silt and clay found in the space between larger sand grains. Matrix usually consists of fine-grained quartz and clay minerals. A matrix-rich sandstone is poorly sorted and often dark in color, which sometimes is called a "dirty sandstones" (Figure 6-1).

Figure 6-1　Sandstones

6　Shale is a fine-grained sedimentary rock notable for its splitting capability. Splitting takes place along the surfaces of very thin layers within the shale. Most shales contain both silt and clay (averaging 2/3 clay-sized clay minerals, 1/3 silt-sized quartz) and are so fine-grained that the surface of the rock feels very smooth. The silt and clay deposits that lithify as shale accumulate on lake bottoms, at the ends of rivers in deltas, beside rivers in flood and on quiet parts of the deep ocean floor[6].

7　Chemical sedimentary rocks are rocks deposited by precipitation of minerals from solution. An example of inorganic precipitation is the formation of rock salt as seawater evaporates.

Chemical precipitation can also be induced by organisms. The sedimentary rock limestone, for instance, can form by the precipitation of calcite within a coral reef by corals and algae[7].

8 Chemical rocks may or may not have once been sediment. Rock salt may form from sediment; individual salt crystals forming in evaporating water act as sediment until they grow large enough to interlock into a solid rock. Minerals that crystallize from solution on the sides of a rock cavity, or as a stalactite[8] in cave, however, were never sediment. Neither was limestone precipitated directly from seawater as a solid rock by corals.

9 Organic sedimentary rocks are rocks that accumulate from the remains of organisms. Coal is an organic rock that forms from the compression of plant remains, such as moss[9], leaves, twigs[10], roots and tree trunks. A limestone formed from the accumulation of clam shells on the sea floor might also be called an organic rock.

10 Different processes of weathering and erosion in different climates result in different types of sediment. The size and nature of the sediments will also be affected by crustal movements in the source area, the distance and type of transportation. The geologists attempt to interpret as much information as possible concerning the nature and environment of the source area, the deposition area and the transporting agents. However, all of this interpretation must be based upon evidence and subject to the strict, nonbiased limitations imposed by the methods of science.

Notes to the text

(1) lithification /ˌlɪθɪfɪˈkeɪʃən/ *n.* 成岩作用，岩化

(2) lithify /ˈlɪθɪfaɪ/ *v.* （使）岩化

(3) cementation /ˌsɪːmenˈteɪʃən/ *n.* 黏结，水泥灌浆；渗碳法

(4) till /tɪl/ *n.* 冰碛

(5) matrix /ˈmeɪtrɪks/ *n.* 基质；脉石

(6) The silt and clay deposits that lithify as shale accumulate on lake bottoms, at the ends of rivers in deltas, beside rivers in flood and on quiet parts of the deep ocean floor.

译文： 泥和黏土通常堆积于湖底、三角洲中的河流末端、洪泛期间的河流旁侧和深海洋底的静海区，它们石化成岩后就转变为页岩。

(7) algae /ˈældʒiː/ *n.* 藻类

(8) stalactite /ˈstæləktaɪt/ *n.* 钟乳石

(9) moss /mɒs/ *n.* 苔藓

(10) twig /twɪg/ *n.* 小枝，末梢；探矿杖

Questions for review

(1) Sedimentary rock has formed from the following except _____.

　　A. lithification　　　　　　　　B. consolidation

　　C. cementation　　　　　　　　D. precipitation

(2) A sedimentary rock that has formed from lithification of sediment is called _____.

　　A. chemical rock　　　　　　　B. clastic rock

　　C. organic rock　　　　　　　　D. inorganic rock

(3) Which is not a chemical or organic sedimentary rock?

　　A. Rock salt.　　　　　　　　　B. Shale.

　　C. Limestone.　　　　　　　　　D. Gypsum.

(4) What is clastic sedimentary rock formed from?

(5) How can conglomerate be distinguished from breccia?

(6) Why haven't angular fragments within breccia moved very far from their source?

(7) How is sandstone formed according to the text?

Unit 7

Economic Geology

Ore Deposit

1 The term "mineral deposits" is used to **denote** a concentration of useful minerals. Mineral deposits include both ores and **nonmetallic** minerals. Scientifically, the word "ore" comprehends all metal-bearing minerals[1] which are commercial sources of the metals, percentages not being considered. Ore is a mineral or an aggregate of minerals from which a valuable constituent, especially a metal, can be profitably mined or extracted. In fact, it is usually considered that, unless a mineral shows enough concentration to repay working, it is not an ore deposit[2]. Technically, an ore is mixed with **barren** matter called "gangue" and capable of being mined at a profit. An ore deposit，an accumulation of ore，is one occurrence of a particular ore type. Most ore deposits are named according to either their location (e.g. the Witswatersrand[3], South Africa), or after a discoverer (e.g. the Kambalda nickel shoots are named after drillers), or after a historical figure, a prominent person, something from **mythology** (phoenix, **kraken**, etc.) or the code name of the resource company which found it (e.g. MKD-5 is the in-house name for the Mount Keith nickel).

2 In this unit, it is with the metallic ores that we are concerned[4]. The common types are gold,

silver, copper, **lead** and iron, which will therefore be used as examples here. Gold mostly occurs as native metal. Silver and copper may also occur native, but in combination with[5] other elements as well. For example, copper occurs along with oxygen as an oxide, or along with sulphur as a **sulphide**. Lead and iron occur in combination with oxygen, sulphur, and carbon as oxides, sulphides, and carbonates. These metals and their ores do not necessarily occur by themselves. Several of them occur and are mined together.

3　Ore deposits can be divided into two large classes. The first class is primary ores that are found in the positions in which they were originally formed, and there are secondary ores which have been transported from their original position by some agency.

4　There are a great many types of primary ore deposits. Some ores have been formed simultaneously with rocks and are actual constituents of igneous rocks. For example, the constituent minerals of an igneous rock **crystallize** out of molten material and during this crystallization minute **grains** of a metal such as iron have flowed together to form a concentrate and so have given rise to workable deposits[6].

5　The secondary ores indicate that although the ore is in its original position, it came into the rock after the rock had been formed, and so was deposited in some kind of cavity or crack. In this case it is obvious that the ores have been deposited from solutions. It is probable that in many cases these solutions were hot and more or less in the condition of gas, but there are many other cases in which the solutions were not hot and in fact were the results of **percolating** rain-water which "**leeched**" the minute grains from the surrounding rocks to concentrate them in some fissure.

6　We know that the rocks which form the crust of the earth have been badly shattered and cracked and cut by many fissures. Rain-water from the surface and hot solutions from the interior of the earth **percolate** along these fissures. Cooling will reduce the capacity of a liquid for carrying material in solution, and so as the gases and liquids pass along the cracks in the rocks they deposit their load of mineral matter, which forms a coating on the walls of the fissures[7]. With successive **encrustations veins** are produced, the ore is not alone but mixed with such "gangue" minerals as quartz and calcite. Unfortunately these commonly occur without ore minerals in them. Such fissure fillings, called fissure veins, are one of the most important kinds of deposit; they are the chief source of gold, silver, copper.

7 Geologists are always searching for more ore deposits to meet constant demand. This has become more difficult with time as easily accessible ore deposits close to the earth's surface have already been exploited by humans in the past. Therefore, more complex techniques have been developed to locate new deposits. However, better and more efficient processing techniques now mean that we can exploit ore deposits which were previously uneconomic. Meanwhile, better extraction techniques are required.

8 The basic extraction of ore deposits follows the steps below:

(1) **Prospecting** or exploration to find and then define the extent and value of ore where it is located ("ore body");

(2) Conducting resource estimation to mathematically estimate the size and grade of the deposit;

(3) Conducting a pre-feasibility study[8] to determine the theoretical economics of the ore deposits, which indicates whether further investment in estimation and engineering studies are warranted and identifies key risks and areas for further work;

(4) Conducting a feasibility study to evaluate the financial viability[9], technical and financial risks and **robustness** of the project and make a decision as whether to develop or walk away from a proposed mine project, which includes mine planning to evaluate the economically recoverable portion of the deposit, the **metallurgy** and ore recoverability, marketability and payability of the ore concentrates, engineering, milling and infrastructure costs, finance and **equity** requirements and a cradle to grave analysis of the possible mine, from the initial **excavation** all the way through to **reclamation**[10];

(5) Development to create access to an ore body and building of mine plant and equipment;

(6) Operation of the mine in an active sense;

(7) Reclamation to make land where a mine had been suitable for future use.

9 Ores (metals) are traded internationally and comprise a sizeable portion of international trade in raw materials both in value and volume. This is because the worldwide distribution of ores is unequal and dislocated from locations of peak demand and from **smelting** infrastructure. Iron ore is traded between customer and producer, though various **benchmark** prices are set quarterly between the major mining conglomerates and the major consumers, and this sets the stage for smaller participants. The World Bank reports that

China was the top importer of ores and metals in 2005 followed by the USA and Japan.

New words

denote /dɪˈnəut/	*vt.*	表示；象征
nonmetallic /ˌnɒnmɪˈtælɪk/	*adj.*	非金属的
	n.	非金属物质
barren /ˈbærən/	*adj.*	无矿的
mythology /mɪˈθɒlədʒɪ/	*n.*	神话；神话学
kraken /ˈkrɑːkən/	*n.*	挪威传说中的北海巨妖
lead /led/	*n.*	矿脉，铅
sulphide /ˈsʌlfaɪd/	*n.*	硫化物
crystallize /ˈkrɪːstəlaɪz/	*vt.*	（使）结晶
grain /greɪn/	*n.*	颗粒；纹理；微量
percolating /ˈpɜːkəleɪtɪŋ/	*n.*	渗透
	adj.	渗透的
	vt.	渗透（percolate 的 ing 形式）
leech /liːtʃ/	*n.*	水蛭；吸血鬼
	vt.&vi.	以水蛭吸血；依附并榨取
percolate /ˈpɜːkəleɪt/	*vt.*	使渗出；使过滤
	vi.	过滤；渗出；渗透
encrustation /ɛnˈkrʌsˈteɪʃən/	*n.*	结壳；积垢
vein /veɪn/	*n.*	［地质］岩脉；纹理；翅脉；性情
prospecting /prəuˈspektɪŋ/	*n.*	探矿；勘探
robustness /rəuˈbʌstnis/	*n.*	稳健性，坚固性
metallurgy /məˈtælədʒɪ/	*n.*	冶金；冶金学；冶金术
equity /ˈekwətɪ/	*n.*	公平；普通股；抵押资产的净值
excavation /ˌekskəˈveɪʃn/	*n.*	挖掘；挖方；矿山巷道；坑道
reclamation /ˌrekləˈmeɪʃən/	*n.*	（废料）回收，再利用；开垦
smelting /ˈsmeltɪŋ/	*n.*	［冶］熔炼；冶炼
	vt.	精炼（smelt 的 ing 形式）
benchmark /ˈbentʃmɑːk/	*n.*	基准，水准点；标准检查程序

Notes to the text

(1) metal-bearing minerals：含金属的矿物

(2) In fact, it is usually considered that, unless a mineral shows enough concentration to repay working, it is not an ore deposit.

译文：实际上，人们普遍认为，除非一种矿物的品位能足以补偿其开采成本，否则不能将之称为矿床。

(3) Witswatersrand：威特沃特斯兰德（南非东北部地区）

(4) In this unit, it is with the metallic ores that we are concerned.

It is…that…是强调句型；we are concerned with 中的 with 被放置在 the metallic ores 之前。

译文：在本单元中，我们所关注的正是金属矿床。

(5) in combination with：与……结合，与……联合

(6) For example, the constituent minerals of an igneous rock crystallize out of molten material and during this crystallization minute grains of a metal such as iron have flowed together to form a concentrate and so have given rise to workable deposits.

give rise to：使发生，引起

译文：比如，构成火成岩的矿物从熔融物结晶出来，在结晶过程中，铁之类的细小金属颗粒汇流于一起富集形成可开发的矿床。

(7) Cooling will reduce the capacity of a liquid for carrying material in solution, and so as the gases and liquids pass along the cracks in the rocks they deposit their load of mineral matter, which forms a coating on the walls of the fissures.

这里的 as 引导时间状语从句。

译文：冷却会降低溶液中液体搬运物质的能力，因此，当气体和液体沿岩石中的裂缝运移时，其所携带的矿物质会在裂缝壁上沉淀，并在裂缝壁上形成一个涂层。

(8) pre-feasibility study：预可行性研究

(9) the financial viability：财务可行性

(10) Conducting a feasibility study to evaluate the financial viability, technical and financial risks and robustness of the project and make a decision as whether to develop or walk away from a proposed mine project, which includes mine planning to evaluate the economically recoverable portion of the deposit, the metallurgy and ore recoverability, marketability and payability of the ore concentrates, engineering,

milling and infrastructure costs, finance and equity requirements and a cradle to grave analysis of the possible mine, from the initial excavation all the way through to reclamation .

a cradle to grave：从生到死；从开始到结束

译文：对它的财务可行性、技术和财务风险、项目的稳定性及对已提出的采矿项目是否继续进行或终止做出决定，其中的采矿设计会对矿床的经济投资回报率、冶金、矿床的恢复、适销、精矿的赢利性、工程、矿石碾磨和基础设施成本、融资和抵押资产的需求进行评估，并对从最初挖掘到回收利用的可能性做出全面分析。

Understand the text

Answer the following questions according to the passage you have read.

(1) What do mineral deposits include?

(2) How should we name the ore deposits?

(3) What are the common types of ores mentioned here?

(4) Do silver and copper occur in combination with other elements?

(5) What are the two large classes of ore deposits?

(6) Where are the secondary ores deposited?

(7) What are the chief sources of gold, silver and copper?

(8) Why is it difficult for geologists to search for more ore deposits nowadays?

(9) Why should geologists conduct a pre-feasibility study about the ore deposits before extraction?

(10) How should geologists conduct a feasibility study to a proposed mine project?

Translation

1. Translate the following sentences into Chinese.

(1) The term "mineral deposits" is used to denote a concentration of useful minerals.

_____

_____

_____

_____

(2) Technically, an ore is mixed with barren matter, called "gangue" and capable of being mined at a profit.

_____

_____

_____

_____

_____

(3) The first class is primary ores that are found in the positions in which they were originally formed, and there are secondary ores which have been transported from their original position by some agency.

_____

_____

_____

_____

_____

(4) We know that the rocks which form the crust of the earth have been badly shattered and cracked and cut by many fissures.

_____

_____

_____

_____

_____

(5) This has become more difficult with time as easily accessible ore deposits close to the earth's surface have already been exploited by humans in the past.

_____

_____

_____

_____

_____

2. Translate the following passage into English.

　　矿床是由地质作用形成的，是由具有开采利用价值的有用矿石构成的集合体。它是地质作用的产物，但又与一般的岩石不同，它具有经济价值。随着科技进步和采矿成本的降低，很多低品位的矿床都已被大量开采。

_____

_____

_____

_____

_____

_____

_____

_____

_____

_____

Writing skill

How to Write an Introduction?

An introduction is the first section of an EST paper, including the background information of the research, the nature and scope of the problem investigated.

It describes the organizational structure of the paper and largely determines the readers' or reviewers' attitude toward this paper.

Generally, the ingredients of an introduction includes three parts: background, problems and outlines of the contents.

- Background
    - introducing the general research field
    - reviewing the previous research

- Problems

  - indicating the problems that have not been solved by previous research

  - proposing schemes & results

- Outlines of the contents

  - giving a description of the main facts and structure of your paper

The following is a sample introduction of Text A :

Nowadays, ores are traded internationally and comprise a sizeable portion of international trade in raw materials both in value and volume. Many countries including China are importing ores for their development. However, easily accessible ore deposits close to the earth's surface have already been exploited. The vital problem is how can we use better and more efficient processing techniques to create greater economic value. Therefore, the purpose of this passage is to introduce the general information of ore deposits and analyze the development of ores trade.

The first part gives the definition of ore deposits; The second part introduces two large classes of ore deposits: primary ores and secondary ores; The third part tells that more complex techniques have been developed to exploit ore deposits, for example: extraction technique.

The end of the passage is the analysis of current situation of international ore trade and some future directions are pointed out as well.

Writing practice

You may try to write an introduction for Text B based on the basic ingredients of an introduction.

Text B

Petroleum

1　The name "petroleum" is derived from the Latin words petra (rock) and oleum (oil), meaning "rock oil". Scientists have proven that most if not all petroleum fields were created by the remains of animal and plant life being compressed on the sea bed by billions of tons of silt and sand several million years ago. When small sea plants and animals die they will sink, then lie on the sea bed where they will decompose[1] and mix with sand and silt. During the decomposition process tiny bacteria[2] will clean the remains of certain chemicals such as phosphorus[3], nitrogen and oxygen. This leaves the remains consisting of mainly carbon and hydrogen. At the bottom of the ocean there is insufficient oxygen for the corpse[4] to decompose entirely. The partially decomposed remains will form a large,gelatinous[5] mass, which will then slowly become covered by multiple layers of sand, silt and mud. This burying process takes millions of years, with layers piling up one atop another. Finally, when the depth of the buried decomposing layer reaches somewhere around 10,000 feet, the natural heat of the earth and the intense pressure will combine to act upon the mass. The end result, over time, is the formation of petroleum. Therefore, created only by nature but exploited by man today, petroleum is considered to be a non-renewable energy source.

2　In its strictest sense, petroleum includes only crude oil. Crude oil is the natural form in which petroleum is first collected, clear, green or black and may be either thin like gasoline or thick like tar[6]. But in common usage it includes all liquid, gaseous, and solid hydrocarbons[7]. Petroleum contains many elements, including carbon (93%-97%), hydrogen (10%-14%), nitrogen (0.1%-2%), oxygen (0.1%-1.5%) and sulphur (0.5%-6%) with a few trace metals making up a very small percentage of its composition. The percentages for these can vary greatly according to geographic region. It is this composition which gives the crude oil its properties.

3　When discussing the composition of petroleum it is worth noticing that the compound of the crude oil tends to dictate[8] the usage of the refined[9] product. Petroleum is generally measured in volume, and for some composition of petroleum it is not cost-effective to refine these into fuel. A lighter, less dense crude oil composition with a compound that contains higher percentages of hydrocarbons is much more profitable as a fuel source. Whereas denser petroleum composition with a less flammable[10] level of hydrocarbons and sulphur are expensive to refine into a fuel and are therefore more suitable for plastics manufacturing and other uses.

4 The petroleum industry generally classifies crude oil by the geographic location it is produced in, its API (American Petroleum Institute) Gravity (an oil industry measure of density), and its sulphur content. Crude oil may be considered light if it has low density or heavy if it has high density; and it may be referred to as sweet if it contains relatively little sulphur or sour if it contains substantial amounts of sulphur.

5 Light crude oil is more desirable than heavy oil since it produces a higher yield of petrol, while sweet oil commands a higher price than sour oil because it has fewer environmental problems and requires less refining to meet sulphur standards imposed on fuels in consuming countries.

6 There are several major oil producing regions around the world. The Kuwait and Saudi Arabia[11]'s oil fields are the largest, although Middle East oil from other countries in the region such as Iran and Iraq also make up a significant part of world production figures.

7 In modern times petroleum is viewed as a valuable commodity, and traded around the world in the same way as gold and diamonds. The petroleum industry is involved in the global processes of exploration, extraction[12], refining, transporting (often with oil tankers[13] and pipelines), and marketing petroleum products. The largest volume products of the industry are fuel oil and petrol.

8 Originally the primary use of petroleum was as a lighting fuel, once it had been distilled[14] and turned into kerosene[15]. When Edison opened the world's first electricity generating plant in 1882 the demand for kerosene began to drop. With the advent[16] of the automobile, gasoline began to be a product in high demand. According to the composition of the crude oil and depending on the demands of the market, refineries[17] can produce different shares of petroleum products. The largest share of oil products is used as "energy carriers", i.e. various grades of fuel oil and gasoline. These fuels include or can be blended to give

gasoline, jet fuel, diesel fuel, heating oil, and heavier fuel oils. Heavier fractions can also be used to produce asphalt, tar, paraffin[18] wax, lubricating and other heavy oils.

9 Petroleum is also a major part of the chemical makeup of many plastics and synthetics[19]. The most startling usage of petroleum for many people is its appearance in foodstuffs such as beer and in medications such as aspirin.

10 Petroleum is vital to many industries, and is of importance to the maintenance of industrialized civilization itself, and thus is critical concern to many nations. Oil accounts for a large percentage of the world's energy consumption. The supply of petroleum is limited, and current estimations tell us that within the next few decades mankind will have completely depleted[20] this valuable natural resource. Although measures have been taken to ensure that there are cheap, renewable fuel options for the eventuality[21] it is still obvious that mankind faces a serious problem when petroleum supplies finally run out.

Notes to the text

(1) decompose /ˌdiːkəmˈpəʊz/ v. 分解，腐烂

(2) bacteria /bækˈtɪərɪə/ n.（复数）细菌

(3) phosphorus /ˈfɒsfərəs/ n. 磷

(4) corpse /kɔːps/ n. 尸体

(5) gelatinous /dʒəˈlætɪnəs/ adj. 凝胶状的

(6) tar /tɑː(r)/ n. 焦油

(7) hydrocarbon /ˌhaɪdrəˈkɑːbən/ n. 碳氢化合物

(8) dictate /dɪkˈteɪt/ v. 命令，指示，支配

(9) refined /rɪˈfaɪnd/ adj. 精炼的；精确的；有教养的

(10) flammable /ˈflæməbl/ adj. 易燃的；可燃性的

(11) Saudi Arabia /ˈsaʊdɪ əˈreɪbɪə/ n. 沙特阿拉伯

(12) extraction /ɪkˈstrækʃn/ n. 抽取，开采

(13) tanker /ˈtæŋkə(r)/ n. 油轮；罐车

(14) distill /dɪsˈtɪl/ v. 蒸馏；提炼

(15) kerosene /ˈkerəsiːn/ n. 煤油

(16) advent /ˈædvənt/ n. 到来；出现

(17) refinery /rɪˈfaɪnərɪ/ n. 精炼厂；炼油厂

(18) paraffin /ˈpærəfɪn/ *n.* 石蜡

(19) synthetic /sɪnˈθetɪk/ *n.* 人工合成物

(20) deplete /dɪˈpliːt/ *v.* 耗尽；使枯竭

(21) eventuality /ɪˌventʃʊˈælətɪ/ *n.* 可能性；不测的事

Questions for review

(1) What created the petroleum fields?

(2) What is the process of petroleum formation?

(3) What does petroleum include in its strictest sense and in common usage?

(4) What elements does petroleum contain?

(5) Which kind of petroleum is more suitable for plastic manufacturing?

(6) How does the petroleum industry generally classify crude oil?

(7) Why is sweet oil more expensive than sour oil?

(8) Which regions around the world are the major oil producing regions?

(9) What was the primary use of petroleum?

(10) What is the largest share of oil products?

Unit 8

Rock Deformation

Folds

1 The movements of the crust are due to two different processes – folding and faulting. If a tablecloth is pushed across a table, it is thrown into a series of ridges and valleys like a sheet of **corrugated** iron[1], and if a part of the earth's crust is squeezed into a smaller space, it is likewise **buckled** into a series of folds. Folds are bends in layered bed rock, which are produced by bending various surfaces, and are the results of **ductile** deformation[2] of rocks. They can occur on all scales and in all environments. Folds often have different shapes and different scales, varying from a few centimeters to hundreds of kilometers[3]. Folds are among the most common tectonic structures[4] found in rocks, and can make some of the most spectacular features.

2 The fact that the rock is folded shows that it was strained plastically rather than by **elastic** or **brittle** strain. Yet the rock exposed in outcrops is generally brittle and shatters when struck with a hammer. The rock is not metamorphosed (most metamorphic rock is intensely folded because it is plastic under the high pressure and temperature environment of deep burial).

Perhaps folding took place when the rock was buried at a moderate depth where high confining pressure[5] favors plastic strain. Alternatively, folding could have taken place close to the surface under a very low rate of strain.

3 Folds are three dimensional structures[6] and therefore, some terms are used to explain the components of folds.

(1) Core: the central part of the fold.

(2) **Limb**: the sides of a fold.

(3) **Hinge**: the line joining points of greatest **curvature** on a folded surface.

(4) **Axial** Plane[7]: the plane or surface that contains hinges or hinge lines on all surfaces involved in folding, which generally divides the fold into nearly equal halves.

(5) Fold profile/Profile section[8]: fold geometry which is always studied in a section **perpendicular** to the hinge line of a fold. In a fold profile, the point of highest elevation on a surface is called **crest** or crestal point[9] and the one of the lowest **trough** or trough point[10].

(6) Inflexion point[11]: the point of least curvature on a folded surface in a fold profile.

4 There are many ways used to classify folds. Folds can be classified by fold shape, tightness, and **dip** or axial plane. According to the shapes of turning point, folds can be roughly classified as follows:

(1) Rounded fold – a fold with a rounded turning point.

(2) Angular fold[12] (**chevron** fold) – a fold with a sharp angular hinge and planar limbs of equal length or a fold that has no curvature in its hinge and straight-sided limbs that form a zigzag pattern.

(3) Box fold[13] – a flat-topped fold with multiple hinges and axial planes.

(4) **Monocline** – a fold with constant regional dip[00] direction.(see diagram below)

5 Lots of different types of folds are grouped based on different criteria. However, according to the bending of single folded layer, folds can be classified into two basic types: **antiform** and **synform**. When the beds are bent upward in an arch, the upward closing fold is called an antiform. When the beds sink into a trough, the downward closing one is called a synform. Folded layers with old rocks in the core are called **anticline,** and those with younger rocks in the core called **syncline**. If the core of an antiform contains the oldest rocks, it is called an antiformal anticline[15]; if the core of a synform contains the youngest

rocks, it is called a synformal syncline[16]. On the other hand, if the core of an antiform contains younger rocks, it is called antiformal syncline[17]; if the core of a synformal structure contains older rocks, it is called synformal anticline[18] (Figure 8-1) .

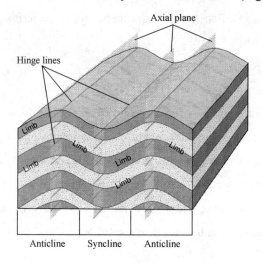

Figure 8-1　Anticline and Syncline

6 Fold tightness is defined by the angle between fold's limbs called interlimb angle[19]. Gentle folds have an interlimb angle of between 180° and 120°; open folds range from 120° to 70°, close folds from 70° to 30° and tight folds from 30° to 0°. **Isoclines**, or isoclinal folds, have an interlimb angle of between 10° and zero, with essentially parallel limbs.

7 Not all folds are equal on both sides of the axis of the fold. Those with limbs of relatively equal length are termed **symmetrical**, and those with highly unequal limbs are **asymmetrical**.

8 We use enveloping surface[20] concept to judge the symmetry of fold: if enveloping surface and axial surface are nearly perpendicular, the fold is said to be symmetrical; otherwise it is asymmetrical. The **vergence** of fold refers to the direction of asymmetric fold. A fold whose axial surface appears to be rotated clockwise relative to the position of axial surface of a symmetric fold is said to have clockwise vergence[21]. Otherwise it has counterclockwise vergence (Figure 8-2).

9 The orientation of fold is decided by the orientations of hinge and axial plane. According to orientation of axial plane, four types of folds can be classified:

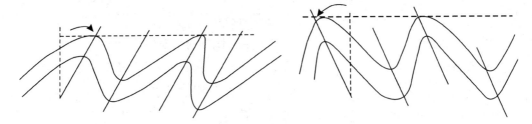

Figure 8-2   Clockwise Vergence (L) and Counterclockwise Vergence (R)

(1) Upright fold[22] often has a nearly vertical axial plane, with two limbs dipping in opposite direction.

(2) Inclined fold[23] has an **oblique** axial plane with two limbs dipping in opposite direction but having different dip angles.

(3) Overturned fold[24] comes into being if the strata on one limb of the fold become nearly upside down.

(4) Recumbent fold[25] refers to an overturned fold with an axial plane that is nearly horizontal.

10 Hinge line may be or may not be linear. The orientation of fold includes trend and plunge. Based on the plunge, fold can be classified as three types:

(1) Horizontal fold[26] – with a horizontal hinge.

(2) Plunge fold[27] – with an inclined hinge.

(3) Vertical fold – with a vertical hinge.

11 The study of folds is very important to show the deformation history of certain regions. Folds are also closely related to the formation of **ores**. At the same time, folds are a very important factor influencing the engineering geological environments. The shapes and patterns of folds reveal much about the nature and extent of the forces of deformation that created them. Familiarity with the geometry of folds helps one understand how to interpret them.

New words

corrugated /ˈkɒrəgeɪtɪd/   *adj.* 波纹的，波状的，褶皱的

buckle /ˈbʌkl/   *vt.* 弯折，屈曲

ductile /ˈdʌktaɪl, -til/   *adj.* 柔软的；延性的，韧性的；易延展的

elastic /ɪˈlæstɪk/   *adj.* 有弹性的；灵活的；易伸缩的

	n. 松紧带；橡皮圈
brittle /ˈbrɪt(ə)l/	*adj.* 易碎的，脆性的；易生气的
limb /lɪm/	*n.* 翼部；翼
hinge /hɪndʒ/	*n.* 枢纽，铰链，转枢
curvature /ˈkɜːvətʃə/	*n.* 曲度，曲率
axial /ˈæksɪəl/	*adj.* 轴的；轴向的
perpendicular /ˌpɜːpənˈdɪkjʊlə(r)/	*adj.* 垂直的，正交的；直立的
crest /krest/	*n.* 山顶；山脊
trough /trɒf/	*n.* 槽；波谷
dip /dɪp/	*n.* 下沉；倾斜；浸渍
chevron /ˈʃevrən/	*n.* 尖顶褶皱
monocline /ˈmɒnəʊklaɪn/	*n.* 单斜层；单斜褶皱
antiform /ˈæntɪfɔːm/	*n.* （地层层序不明的）背斜构造
synform /ˈsɪnfɔːm/	*n.* 向形，向斜式构造
anticline /ˈæntɪklaɪn/	*n.* 背斜；背斜层
syncline /ˈsɪŋklaɪn/	*n.* 向斜；向斜层
isocline /ˈaɪsəʊklaɪn/	*n.* [地物] 等斜线；等斜褶皱
symmetrical /sɪˈmetrɪkəl/	*adj.* 匀称的，对称的
asymmetrical /ˌeɪsɪˈmetrɪkəl/	*adj.* 非对称的；不均匀的
vergence /ˈvɜːdʒəns/	*n.* 趋异，构造转向；朝向
oblique /əˈbliːk/	*n.* 倾斜物
	adj. 斜的；不光明正大的
	vi. 倾斜
ore /ɔː(r)/	*n.* 矿；矿石

Notes to the text

(1) corrugated iron：波状铁，瓦楞铁

(2) ductile deformation：塑/延性变形

(3) Folds often have different shapes and different scales, varying from a few centimeters to hundreds of kilometers.

varying from 是现在分词做伴随状况状语。

译文：褶皱的形状和规模大多不一，从几厘米到几百千米不等。

(4) tectonic structures：地质构造

(5) confining pressure：围压；封闭压力

(6) three dimensional structures：三维结构

(7) axial plane：轴面

(8) profile section：剖面图；纵剖面

(9) crestal point：脊点

(10) trough point：槽点；波谷

(11) inflexion point：拐点；转折点

(12) angular fold：尖棱褶皱

(13) box fold：箱状褶皱

(14) regional dip：区域倾斜

(15) antiformal anticline：背形背斜

(16) synformal syncline：向形向斜

(17) antiformal syncline：背形向斜

(18) synformal anticlines：向形背斜

(19) interlimb angle：翼间角；两翼之间的内夹角

(20) enveloping surface：包络面；褶皱包围面

(21) A fold whose axial surface appears to be rotated clockwise relative to the position of axial surface of a symmetric fold is said to have clockwise vergence.

该句的主干结构是 A fold is said to have clockwise vergence. 其中，whose…是 A fold 的定语从句，在该从句中，relative to…是形容词短语做 clockwise 的后置定语。

译文：相对于对称褶皱的轴面位置，轴面按顺时针方向旋转的不对称褶皱被认为是顺时针转向。

(22) upright fold：直立褶皱

(23) inclined fold：倾斜褶皱

(24) overturned fold：倒转褶皱

(25) recumbent fold：伏褶皱；平/伏卧褶皱

(26) horizontal fold：水平褶皱；不倾覆褶皱

(27) plunge fold：倾伏褶皱

Understand the text

Answer the following questions according to the passage you have read.

(1) How do you describe the components of folds?

(2) What ways can be used to classify the folds mentioned here?

(3) How to distinguish the antiformal anticline from the antiformal syncline?

(4) What is interlimb angle? How to define different folds based on interlimb angle?

(5) How do we judge the symmetry of fold?

(6) How many types of folds can be classified according to orientation of axial plane? What are they?

(7) Why is it important to study the folds?

Translation

1. Translate the following sentences into Chinese.

(1) Folds are among the most common tectonic structures found in rocks, and can make some of the most spectacular features.

_____

_____

_____

_____

(2) The line joining points of greatest curvature on a folded surface is called the hinge of a fold.

_____

_____

_____

_____

(3) The fact that the rock is folded shows that it was strained plastically rather than by elastic or brittle strain.

_____

_____

(4) On the other hand, if the core of an antiform contains younger rocks, it is called antiformal syncline; if the core of a synformal structure contains older rocks, it is called synformal anticline.

(5) Upright fold often has a nearly vertical axial plane, with two limbs dipping in opposite direction.

2. Translate the following passage into English.

在组成地壳的构造板块中，其中两个板块沿它们的边界互相挤压便形成了褶皱山系。巨大的压力迫使板块的边缘发生弯曲和向上隆起形成一系列的褶皱。褶皱山系通过所谓的"造山运动"而形成。因为板块每年只位移几厘米，因而一次造山运动形成一个山脉需要数百万年。褶皱山系是地球上最常见的山脉类型。其他类型有火山山脉、侵蚀山和断块山。火山活动可形成火山山脉。侵蚀山的产生是由于风和水剥蚀掉了陆地中松软的部分而留下了坚硬的岩丘。断块山形成于大陆地壳发生位移的部位。

Writing skill

How to Write a Summary?

For academic papers of EST, a summary is a shortened version of the original. It gives the main facts or ideas of the paper, but not the details. A good summary is a bridge between readers and the author: for readers, it enables them to understand the overview quickly; for the author, it is the best way to demonstrate what he is writing in this paper.

A summary should be brief, complete, accurate, coherent and objective.

● Brief: to omit unnecessary details like examples, explanations and other unimportant information. The length of a summary should be one-third or one-quarter of its original size.

● Complete: to contain all the main and supporting points

● Accurate: to give the same attention and stress to the points as the author does.

● Coherent: to link the context by necessary transitions and function structures.

Usually, the following transition words are used in a summary:

For contrast: *in spite of, on the one hand,...on the other hand,..., nevertheless, nonetheless, notwithstanding, in/by contrast, on the contrary,...*

For sequence or order: *first, second, third, next, then, finally,...*

For cause and effect: *accordingly, consequently, hence, therefore, due to, for this reason, as a result,...*

For additional support: *besides, equally important, further, furthermore, in addition, moreover, ...*

For conclusion: *finally, in a word, in brief, in conclusion, in the* end, *on the whole, in sum,...*

● Objective: to reflect the content of the original passage objectively, not including your own ideas or emotions on the topic.

The sample summary of the following paragraphs taken from Text A is written for your reference:

The source paragraphs:

"Lots of different types of folds are grouped based on different criteria. According to the bending of single folded layer, folds can be classified into two basic types: antiform and synform. When the beds are bent upward in an arch, the upward closing fold is called an antiform. When the beds sink into a trough, the downward closing one is called a synform. Folded layers with old rocks in the core are called anticline, and those with younger rocks in the core called syncline. If the core of an antiform contains the oldest rocks, it is called an antiformal anticline(15); if the core of a synform contains the youngest rocks, it is called a synformal syncline. On the other hand, if the core of an antiform contains younger rocks, it is called antiformal syncline; if the core of a synformal structure contains older rocks, it is called synformal anticline.

Fold tightness is defined by the angle between fold's limbs called interlimb angle. Gentle folds have an interlimb angle of between 180° and 120°; open folds range from 120° to 70°, close folds from 70° to 30° and tight folds from 30° to 0°. Isoclines, or isoclinal folds, have an interlimb angle of between 10° and 0°, with essentially parallel limbs."

The sample summary:

Folds are classified into different types according to different criteria: the bending of single folded layer and the rocks in the core of folded layers.

Different interlimb angles define the fold tightness as gentle golds, open folds, close folds, tight folds and isoclinal folds.

Writing practice

You may try to write a summary based on Text A.

Text B

Faults

1 Our interest in faults is practical as well as scientific and aesthetic. Understanding faults is useful, especially because of the devastating effects of active faults on populations who live near them, in the design for long-term stability of dams, bridges, buildings, and power plants[(1)]. The study of faults helps us to understand mountain-building and deformation processes[(2)], and also yields results of practical value.

2 The continents we live on are parts of moving plates. Most of the action takes place where plates meet. Plates may collide, pull apart, or scrape past each other. All the stress and strain produced by moving plates build up in the earth's rocky crust until it simply can't take it anymore and all at once, CRACKS! The rock breaks and the two rocky blocks move toward opposite directions along a more or less planar fracture surface called fault.

3 Faults have produced some of the earth's most spectacular scenery. In geology, a fault is a planar fracture or discontinuity in a volume of rocks, across which there has been significant displacement along the fractures as a result of the earth movement[(3)]. Large faults within the earth's crust result from the action of plate tectonic forces[(4)], such as subduction zones[(5)] or transform faults[(6)]. Energy release associated with rapid movement on active faults is the cause of most earthquakes[(7)]. Active faults, though they may not move for decades, can move many feet in a matter of seconds, producing an earthquake. Faults occur in many forms and dimensions. They may be hundreds of kilometers or only a few centimeters long. Their outcrop traces may be straight or sinuous[(8)]. They may occur as knife-sharp boundaries or as fault or shear zones[(9)], with millimeters to several kilometers thick. The most obvious feature related to faulting is the displacement of some markers, most commonly bedding[(10)]. Displacement occurs along the actual movement surface – the fault "plane" – commonly non-planar. If the fault plane is not vertical, the rock mass[(11)] resting on the fault plane is called hanging wall[(12)], and the rock mass beneath the fault plane is called footwall[(13)].

4 We can measure the dip and strike of fault plane[14] and make other measurements of structures, such as slickensides[15], which indicates relative movement along the fault plane.

5 The two classes of faults include dip-slip[16] (up and down movement), which is further divided into normal and thrust (reverse) faults; and strike-slip[17] (movement parallel to the fault plane). A fault in which the hanging wall moves down and the footwall is stationary is called normal fault[18]. Normal faults are formed by tensional, or pull-apart forces. Normal faults occur mainly in areas where the crust is being extended such as a divergent boundary[19]. A fault in which the hanging wall is the upthrown side is called thrust fault[20] because the hanging wall appears to have been pushed up over the footwall. Such faults are formed by compressional forces that push rock together and are by far the most common of dip-slip faults. Thrust faults occur in areas where the crust is being shortened such as at a convergent boundary[21].

6 A strike-slip fault is a surface along which one rock mass has moved horizontally with respect to[22] the other. A strike-slip fault is the result of oppositely-directed forces in the crust that do not act along the same line, so that it is a distortion[23] of the crust rather than a change in area. Transform boundaries[24] are a particular type of strike-slip fault. Many earthquakes are caused by movement on faults that have components of both dip-slip and strike-slip which is known as oblique slip[25]. The largest earthquakes may occur along thrust faults. For example, the Longmenshan Fault is a thrust fault which runs along the base of Longmen Mountains in Sichuan basin in southwestern China and along which the 2008 Wenchuan earthquake occurred(Figure 8-3).

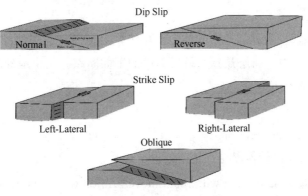

Figure 8-3   Classification of Faults

7 In real life, faulting is not such a simple picture! Usually faults do not have purely up-and-down or side-by-side movement as described above. It's much more common to have some combination of fault movements occurring together.

8 In geotechnical engineering[26], a fault often forms a discontinuity that may have a large influence on mechanical behavior (strength, deformation, etc.) of soil and rock masses in, for example, tunnel, foundation, or slope construction. The level of a fault's activity can be critical for locating buildings, tanks, and pipelines and assessing seismic[27] shaking and tsunami[28] hazard to infrastructure[29] and people in the vicinity[30]. Radiocarbon dating[31] of organic material buried next to or over a fault shear is often critical in distinguishing active from inactive faults. From such relationships, paleoseismologists[32] can estimate the sizes of past earthquakes over the past several hundred years, and develop rough projections of future fault activity.

Notes to the text

(1) Understanding faults is useful, especially because of the devastating effects of active faults on populations who live near them, in the design for long-term stability of dams, bridges, buildings, and power plants.

译文：活跃的断层带会给居住在其附近的人们带来灾难性的影响。因此，了解断层非常有用，可以帮助人们设计出结实耐用的堤坝、桥梁、建筑物和发电厂。

(2) deformation process：变形过程，变形作用

(3) In geology, a fault is a planar fracture or discontinuity in a volume of rocks, across which there has been significant displacement along the fractures as a result of the earth movement.

介词+关系代词引导的非限制性定语从句。which 指代上文中的"fault"。

译文：在地质学中，断层是指大量岩石中的平面断裂。由于地球的运动，导致断裂发生明显的位置偏移。

(4) plate tectonic force：板块构造力

(5) subduction zone：俯冲带

(6) transform fault：转换断层；转形断层

(7) Energy release associated with rapid movement on active faults is the cause of most earthquakes.

该句的主干结构是 Energy release is the cause of most earthquakes; associated with…做 Energy release 的定语。

active faults：活动断裂；活性断层。

译文：与活动断层的快速运动相关的能量释放是大多数地震的诱因。

(8) sinuous /ˈsɪnjʊəs/ *adj.* 弯曲的；波折的；蜿蜒的；迂回的

(9) shear zone：（地质）剪碎带

(10) bedding /ˈbedɪŋ/ *n.* 层理；成层；基底

(11) rock mass：岩体；岩块；岩石层

(12) hanging wall：上盘；顶壁

(13) footwall：下盘；底壁

(14) the dip and strike of the fault plane：断层面的倾向和走向

(15) slickenside：擦痕面，断层擦面

(16) dip-slip fault：倾向滑动断层

(17) strike-slip fault：走向滑动断层

(18) normal fault：正断层

(19) divergent boundary：离散边界

(20) thrust fault：逆断层；冲断层

(21) convergent boundary：会聚边界，聚合边界

(22) with respect to：相对于

(23) distortion /dɪˈstɔːʃn/ *n.* 变形；扭变，畸变

(24) transform boundary：转换边缘

(25) oblique slip　斜向滑动

(26) geotechnical engineering：岩土工程

(27) seismic /ˈsaɪzmɪkˈ/ *adj.* 地震的

(28) tsunami /tsuːˈnɑːmɪ/ *n.* 海啸

(29) infrastructure /ˈɪnfrəstrʌktʃə(r)/ *n.* 深部构造，下构造层；基础设施

(30) in the vicinity：在附近

(31) radiocarbon dating：碳同位素年龄测定

(32) paleoseismologist /ˌpeɪlɪəusaɪzˈmɒlədʒɪst/ *n.* 古地震学家

Questions for review

(1) Normal and reverse faults are examples of _____.

　　A. strike-slip

　　B. dip-slip

　　C. oblique-slip

　　D. down-slip

(2) An inclined surface along which a rock mass slipped downward, is a _____.

　　A. normal fault

　　B. thrust fault

　　C. strike-slip fault

　　D. oblique-slip fault

(3) _____ are grouped into strike-slip faults.

　　A. Divergent boundaries

　　B. Convergent boundaries

　　C. Transform boundaries

　　D. Subducting plate boundaries

(4) The largest earthquakes often occur along _____.

　　A. normal faults

　　B. thrust faults

　　C. strike-slip faults

　　D. oblique-slip faults

(5) Which one of the following is correct according to this passage?

　　A. The outcrop traces of faults are straight.

　　B. The rock mass resting on the fault plane is called the footwall.

　　C. Normal faults are formed by compressional forces.

　　D. Energy release associated with rapid movement on active faults may probably cause earthquakes.

(6) How do dip-slip faults differ from strike-slip faults?

(7) What kinds of faults are in the following figures respectively?

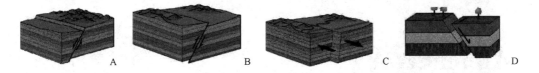

(8) What influences do the faults have on structures and people?

Appendix A

Glossary

A

aragonite /ˈərægənaɪt/ *n.*［矿物］霰石　　　　　　　　　　　　　4-A

arch /ɑ:tʃ/ *n.* 弓形，拱形；拱门　*vt.* 使……弯成弓形；用拱连接

　　　　　vi. 拱起；成为弓形　　　　　　　　　　　　　　　　3-A

argon /ˈɑ:gɒn/ *n.*［化］氩（18 号化学元素，符号为 Ar）　　　1-A

artificial /ˌɑ:tɪˈfɪʃl/ *adj.* 人造的　　　　　　　　　　　　　　　1-A

astounding /əsˈtaundɪŋ/ *adj.* 令人震惊的；令人惊骇的　　　　　2-A

asymmetrical /ˌeɪsɪˈmɛtrɪkəl/ *adj.* 非对称的；不均匀的　　　　8-A

attrition /əˈtrɪʃn/ *n.* 摩擦；磨损　　　　　　　　　　　　　　3-B

augite /ˈɔ:dʒaɪt/ *n.* 普通辉石；斜辉石　　　　　　　　　　　　4-B

axial /ˈæksɪəl/ *adj.* 轴的；轴向的　　　　　　　　　　　　　　8-A

B

bacteria /bækˈtɪərɪə/ *n.*（复数）细菌　　　　　　　　　　　　7-B

barren /ˈbærən/ *adj.* 无矿的　　　　　　　　　　　　　　　　　7-A

basalt /ˈbæsɔ:lt/ *n.*［岩］玄武岩；黑陶器　　　　　　　　　　5-A

bedding /ˈbedɪŋ/ *n.* 层理；成层；基底　　　　　　　　　　　　8-B

bedload /ˈbedləud/ *n.* 推移质　　　　　　　　　　　　　　　　3-B

benchmark /ˈbentʃmɑ:k/ *n.* 基准，水准点；标准检查程序　　　7-A

binder /ˈbaɪndə(r)/ *n.* 黏合剂；活页夹；煤层杂质条纹　　　　6-A

biotite /ˈbaɪətaɪt/ *n.* 黑云母　　　　　　　　　　　　　　　　4-B

boulder /ˈbəuldə(r)/ *n.* 巨砾，漂砾　　　　　　　　　　　　　1-B

brittle/ˈbrɪt(ə)l/ *adj.* 易碎的，脆性的；易生气的　　　　　　　8-A

buckle /ˈbʌkl/ *vt.* 弯折，弯曲　　　　　　　　　　　　　　　8-A

buckling /ˈbʌklɪŋ/ *n.* 弯折，挤弯作用　　　　　　　　　　　　2-A

buff /bʌf/ *vt.* 软皮摩擦；缓冲；擦亮，抛光某物　　　　　　　3-A

C

calcite /ˈkælsaɪt/ *n.* 方解石　　　　　　　　　　　　　　　　4-A

cavitation /ˌkævɪˈteɪʃ(ə)n/ *n.* 成穴；空化作用；气蚀；空穴现象　3-B

cavity /ˈkævətɪ/ *n.* 洞；空穴　　　　　　　　　　　　　　　　4-A

cementation /ˌsɪ:menˈteɪʃən/ *n.* 黏结，水泥灌浆；渗透法　　　6-B

Cenozoic /ˌsiːnəˈzəʊɪk/ *n.*（等于 Cainozoic）新生代　　　　　　　　　2-B

certify /ˈsɜːtɪfaɪ/ *vt.* 证明；发证书给……　　　　　　　　　2-A

chevron /ˈʃevrən/ *n.* 尖顶褶皱　　　　　　　　　8-A

chisel /ˈtʃɪzl/ *vt.* 雕，刻；凿；欺骗 *vi.* 雕，刻；凿；欺骗 *n.* 凿子　　　3-A

chlorine /ˈklɔːriːn/ *n.*［化］氯（17 号化学元素，符号为 Cl）　　　4-A

cinder /ˈsɪndə(r)/ *n.* 火山渣；炉渣　　　　　　　　　2-A

circumference /səˈkʌmf(ə)r(ə)ns/ *n.* 圆周，周围　　　　　　　　　1-A

clam /klæm/ *n.* 蛤蜊　　　　　　　　　4-A

cleavage /ˈkliːvɪdʒ/ *n.*［矿物］解理；劈理　　　　　　　　　4-A

coalesce /ˌkəʊəˈles/ *v.* 使联合；合并　　　　　　　　　5-B

conceal /kənˈsiːl/ *vt.* 隐藏；隐瞒；掩盖　　　　　　　　　2-A

concentration /kɒnsnˈtreɪʃn/ *n.* 浓度；集中；浓缩　　　　　　　　　4-A

conglomerate /kənˈɡlɒm(ə)rət/ *n.* 企业集团，大型联合企业　　　7-A

conglomerate /kənˈɡlɒmərət/ *n.*［岩］砾岩；聚合物 *vt.* 使聚结；凝聚成团

　　　　　　　　　　　　vi. 凝聚成团　　　　　　　　　6-A

constituent /kənˈstɪtjʊənt/ *n.* 成分，组成部分 *adj.* 构成的；选举的　　　4-A

contorted /kənˈtɔːtid/ *adj.* 弯曲的，扭曲的 *v.* 扭曲；歪曲（contort 的过去分词）　　　2-A

coral /ˈkɒrəl/ *n.* 珊瑚；珊瑚虫 *adj.* 珊瑚的；珊瑚色的　　　4-A

core /kɔː/ *n.*［地］地核；岩心　　　　　　　　　1-A

corpse /kɔːps/ *n.* 尸体　　　　　　　　　7-B

corrosion /kəˈrəʊʒn/ *n.* 腐蚀　　　　　　　　　3-B

corrugated /ˈkɒrəɡeɪtid/ *adj.* 波纹的，波状的，褶皱的　　　8-A

crest /krest/ *n.* 山顶；山脊　　　　　　　　　8-A

crevice /ˈkrevɪs/ *n.* 裂缝；裂隙　　　　　　　　　3-A

crumble /ˈkrʌmbl/ *vi.* 崩溃；破碎，崩解 *vt.* 崩溃；弄碎，粉碎　　　3-A

crust /krʌst/ *n.* 地壳；外壳，坚硬的外壳，面包皮　　　1-A

crystal /ˈkrɪstəl/ *n.* 水晶，晶体　　　　　　　　　1-B

crystalline /ˈkrɪstəlaɪn/ *adj.* 结晶的；水晶般的　　　　　　　　　4-A

crystallize /ˈkrɪːstəlaɪz/ *v.*（使）结晶　　　　　　　　　7-A

curvature /ˈkɜːvətʃə/ *n.* 曲度，曲率　　　　　　　　　8-A

D

debris /ˈdeɪbrɪ:/ *n.* 岩屑，碎石，尾矿		2-A
decipher /dɪˈsaɪfə(r)/ *n.* 密电译文 *vt.* 破译（密码）；解释		2-A
decompose /ˌdi:kəmˈpəʊz/ *v.* 分解，腐烂		7-B
decomposed /ˌdi:kəmˈpəʊzd/ *adj.* 已腐烂的，已分解的		3-A
decomposition /ˈdi:kɒmpəˈzɪʃən/ *n.* 分解，腐烂；变质		3-A
delta /ˈdeltə/ *n.*（河流的）三角洲		3-B
denote /dɪˈnəut/ *vt.* 表示；象征		7-A
depict /dɪˈpɪkt/ *vt.* 描述；描画		2-A
deplete /dɪˈpli:t/ *vt.* 耗尽；使枯竭		7-B
designate /ˈdezɪgneɪt/ *vt.* 指明，指出；指派		2-A
detrital /dɪˈtraɪtəl/ *adj.* 碎屑的；由岩屑形成的		6-A
diagnostic /daɪəgˈnɒstɪk/ *n.* 诊断法；诊断结论 *adj.* 诊断的；特征的		4-A
diameter /daɪˈæmɪtə/ *n.* 直径		1-A
dictate /dɪkˈteɪt/ *vt.* 命令，指示，支配		7-B
dioxide /daɪˈɒksaɪd/ *n.* 二氧化物		1-A
dip /dɪp/ *n.* 下沉；倾斜；浸渍		8-A
distill /dɪsˈtɪl/ *v.* 蒸馏；提炼		7-B
distortion /disˈtɔ:ʃən/ *n.* 变形；扭曲，扭变，畸变		8-B
disturbance /dɪˈstɜ:bəns/ *n.* 困扰；动乱；干扰；妨碍		2-A
ductile /ˈdʌktaɪl/ *adj.* 柔软的；延性的，韧性的；易延展的		8-A
dune /dju:n/ *n.*（由风吹积而成的）沙丘		3-A
dynamic /daɪˈnæmik/ *adj.* 动态的；充满活力的，精力充沛的；不断变化的，充满变数的		1-A

E

elastic /ɪˈlæstɪk/ *adj.* 有弹性的；灵活的；易伸缩的 *n.* 松紧带；橡皮圈		8-A
emerald /ˈemərəld/ *n.* 祖母绿；绿宝石；翡翠		4-B
encroachment /ɪnˈkrɒtʃmənt/ *n.* 侵入，侵蚀		2-A
encrustation /ɛnˈkrʌsˈteɪʃən/ *n.* 结壳；积垢		7-A

episode /ˈepɪsəud/ *n.* 一段经历，插曲，片段 　　　　　　　　　　2-A

equator /iˈkweitə/ *n.* 赤道 　　　　　　　　　　　　　　　　　　　1-A

equity /ˈekwətɪ/ *n.* 公平；普通股；抵押资产的净值 　　　　　　　　7-A

erosion /ɪˈrəʊʒn/ *n.* 侵蚀，冲刷 　　　　　　　　　　　　　　　　2-A

evaporate /ɪˈvæpəreɪt/ *vt.* 蒸发；使……脱水；使……消失 *vi.* 蒸发，挥发；消失，

　　　　　　失踪 　　　　　　　　　　　　　　　　　　　　　　3-A

eventuality /ɪˌventʃuˈælətɪ/ *n.* 可能性；不测的事 　　　　　　　　7-B

excavation /ˌekskəˈveɪʃn/ *n.* 挖掘；挖方；矿山巷道；坑道 　　　　7-A

exogenetic /ˈeksəudʒɪˈnetɪk/ *adj.* 外生的；外因的；外源性的 　　　3-A

extraction /ɪkˈstrækʃn/ *n.* 抽取，开采 　　　　　　　　　　　　　7-B

extrusive /ɪkˈstruːsɪv/ *adj.* 突出的；喷出的，挤出的 　　　　　　　5-A

F

feldspar /ˈfeldspɑː/ *n.* 长石 　　　　　　　　　　　　　　　　　　4-B

feldspathoid /ˈfeldspæθɔɪd/ *n.* 似长石 *adj.* 似长石的 　　　　　　5-A

fissure /ˈfɪʃə(r)/ *n.* 裂缝；裂沟（尤指岩石上的） 　　　　　　　　3-B

fizz /fiz/ *n.* 发嘶嘶声；起泡沫 　　　　　　　　　　　　　　　　　4-B

flammable /ˈflæməbl/ *adj.* 易燃的；可燃性的 　　　　　　　　　　7-B

flattened /ˈflætnd/ *adj.* 扁平的；平缓的 *v.* 把……弄平；变平（flatten 的过去分词）1-A

floodplain /ˈflʌdpleɪn/ *n.* 泛滥平原，漫滩，洪积平原 　　　　　　3-B

fluvial /ˈfluːvɪəl/ *adj.* 河流的；冲积的 　　　　　　　　　　　　　3-B

footwall /ˈfʊtwɔːl/ *n.* 下盘；底壁 　　　　　　　　　　　　　　　8-B

formula /ˈfɔːmjələ/ *n.* ［数］公式，准则；配方 　　　　　　　　　4-A

friction /ˈfrɪkʃn/ *n.* 摩擦；摩擦力 　　　　　　　　　　　　　　　5-B

frigid /ˈfrɪdʒɪd/ *adj.* 寒冷的，严寒的 　　　　　　　　　　　　　3-A

G

gangue /gæŋ/ *n.* ［地质］脉石；矿石；尾矿 　　　　　　　　　　4-A

garnet /ˈgɑːnɪt/ *n.* ［矿物］石榴石；深红色 *adj.* 深红色的；暗红色的 4-A

gelatinous /dʒəˈlætɪnəs/ *adj.* 凝胶状的 　　　　　　　　　　　　7-B

geochemistry /dʒiːəʊˈkemɪstrɪ/ *n.* 地球化学 　　　　　　　　　1-B

geological /dʒɪəˈlɑdʒɪkl/ *adj.* 地质（学）的   1-A

geologist /dʒɪˈɒlədʒɪst/ *n.* 地质学家   1-A

geophysics /ˈdʒiːəʊˈfɪzɪks/ *n.* 地球物理学   1-B

glacier /ˈɡlæsɪə(r)/ *n.* 冰河，冰川   1-B

gneiss /naɪs/ *n.* 片麻岩   5-B

grain /ɡreɪn/ *n.* 颗粒；纹理；微量   7-A

granite /ˈɡrænɪt/ *n.* 花岗岩；花岗石   5-A

gravel /ˈɡræv(ə)l/ *n.* 沙砾，碎石   2-A

groove /ɡruːv/ *n.* 沟；槽 *vt.* 开槽于 *vi.* 形成沟槽   4-A

H

halite /ˈhælaɪt/ *n.* 岩盐；石盐   4-A

haze /heɪz/ *n.* 雾团，霾   2-A

heaving /ˈhiːvɪŋ/ *n.* 冻胀，隆起   2-A

herbivore /ˈhɜːbɪvɔː/ *n.* 草食动物   2-B

highlight /ˈhaɪlaɪt/ *vt.* 强调；照亮；使突出

                n. 加亮区；精彩部分；最重要的细节或事件；闪光点   2-A

hinge /hɪndʒ/ *n.* 枢纽，铰链，转枢   8-A

homogeneous /ˌhɒməˈdʒiːnɪəs/ *adj.* 均质的；均匀的   4-A

hornblende /ˈhɔːnblend/ *n.* （普通）角闪石   4-B

hospitable /hɒˈspɪtəbəl/ *adj.* 好客的；热情友好的；（环境）舒适的   1-A

hydrocarbon /ˌhaɪdrəˈkɑːbən/ *n.* 碳氢化合物   7-B

hydrosphere /ˈhaɪdrəsfɪə/ *n.* 水圈，水界，水气   1-A

hypabyssal /ˌhɪpəˈbɪsəl/ *adj.* 半深成的，浅成的   5-A

I

igneous /ˈɪɡnɪəs/ *adj.* 火的；[岩] 火成的；似火的   5-A

impart /ɪmˈpɑːt/ *vt.* 给予，传授；告知，透露   6-A

implode /ɪmˈpləʊd/ *vt.* 向内破裂；内爆；突然崩溃；向内聚爆；向内坍塌   3-B

index /ˈɪndeks/ *n.* 指标；指数；索引；指针 *vt.* 指出；编入索引 *vi.* 做索引   2-A

infrastructure /ˈɪnfrəˌstrʌktʃə(r)/ *n.* 深部构造，下构造层；基础设施   8-B inorganic /ˌɪnɔːˈgænɪk/ *adj.* 无机的；无生物的 　　　　　　　　　4-A

interaction /ˌɪntərˈækʃən/ *n.* 相互影响；相互作用 　　　　　　　1-A

interior /ɪnˈtɪəriə(r)/ *n.* 内部，内陆

　　　　　　　　adj. 内部的，本质的；国内的 　　　　　　　　　1-A

interlock /ˌɪntəˈlɒk/ *v.* 互锁；连锁 　　　　　　　　　　　　　5-B

interrelated /ˌɪntərɪˈleɪtɪd/ *adj.* 相互关联的

　　　　　　　　v. 相互关联［影响］（interrelate 的过去式和过去分词） 1-A

intrusive /ɪnˈtruːsɪv/ *adj.* 侵入的 　　　　　　　　　　　　　5-A

invertebrate /ɪnˈvɜːtɪbrət, -breɪt/ *n.* 无脊椎动物 　　　　　　2-B

ion /ˈaɪən/ *n.* 离子 　　　　　　　　　　　　　　　　　　　5-B

isocline /ˈaɪsəʊklaɪn/ *n.* ［地物］等斜线；等斜褶皱 　　　　　8-A

K

karst /kɑːst/ *n.* 喀斯特地形（石灰岩地区常见的地形）；岩溶 　3-A

kerosene /ˈkerəsiːn/ *n.* 煤油 　　　　　　　　　　　　　　7-B

kraken /ˈkrɑːkən/ *n.* 挪威传说中的北海巨妖 　　　　　　　　7-A

L

lava /ˈlɑːvə/ *n.* 熔岩；火山岩 　　　　　　　　　　　　　　2-A

lead /led/ *n.* 矿脉，铅 　　　　　　　　　　　　　　　　　7-A

leech /liːtʃ/ *n.* 水蛭；吸血鬼 *vt.&vi.* 以水蛭吸血；依附并榨取 7-A

limb /lɪm/ *n.* 翼部；翼 　　　　　　　　　　　　　　　　　8-A

limestone /ˈlaɪmstəʊn/ *n.* 石灰岩 　　　　　　　　　　　　2-A

lithification /ˌlɪθɪfɪˈkeɪʃən/ *n.* 成岩作用，岩化 　　　　　　　6-B

lithify /ˈlɪθɪfaɪ/ *v.* （使）岩化 　　　　　　　　　　　　　6-B

lithosphere /ˈlɪθəsfɪə/ *n.* ［地物］［地质］岩石圈；陆界 　　　1-A

lore /lɔː(r)/ *n.* 学问；知识；传说 　　　　　　　　　　　　　2-A

M

magma /ˈmægmə/ *n.* ［地质］岩浆；糊剂 　　　　　　　　　5-A

magnesium /mægˈniːziəm/ *n.* 镁 　　　　　　　　　　　　4-B

mantle /ˈmæntl/ *n.* 地幔；罩；盖层 　　　　　　　　　　　1-A

matrix /ˈmeɪtrɪks/ *n.* 基质；脉石　　　　　　　　　　　　　　　　6-B

meander /miˈændə/ *n.* 曲流（常用复数）；河曲　　　　　　　　　　3-B

metallurgy /məˈtælədʒɪ/ *n.* 冶金；冶金学；冶金术　　　　　　　　7-A

metamorphic /metəˈmɔːfɪk/ *adj.* 变质的；变形的　　　　　　　　　5-A

metamorphism /metəˈmɔːfɪz(ə)m/ *n.* 变质；变性　　　　　　　　5-B

metamorphose /ˌmetəˈmɔːfəuz/ *v.* 变质；变形　　　　　　　　　5-B

meteorology /ˌmiːtɪəˈrɒlədʒi/ *n.* 气象学，气象状态　　　　　　　1-B

mica /ˈmaikə/ *n.* 云母　　　　　　　　　　　　　　　　　　　1-B

minute /maiˈnjuːt/ *adj.* 微小的；详细的；细致的；精密的　　　　　4-A

molecule /ˈmɒlɪkjuːl/ *n.* 分子，克分子　　　　　　　　　　　　1-B

molten /ˈməultən/ *adj.* 熔化的　*v.* 熔化，溶解，变软　　　　　　1-A

monocline /ˈmɒnəuklaɪn/ *n.* 单斜层；单斜褶皱　　　　　　　　　8-A

moraine /məˈreɪn/ *n.* 冰碛；（熔岩流表面的）火山碎屑　　　　　　3-A

moss /mɒs/ *n.* 苔藓　　　　　　　　　　　　　　　　　　　　6-B

muscovite /ˈmʌskəuvaɪt/ *n.* 白云母　　　　　　　　　　　　　4-B

mysticism /ˈmɪstəˌsɪzəm/ *n.* 神秘；神秘主义；谬论　　　　　　　2-A

mythology /mɪˈθɒlədʒi/ *n.* 神话；神话学　　　　　　　　　　　7-A

N

niche /niːʃ, nɪtʃ/ *n.* ［生］生态龛；生态位；小生态环境　　　　　2-B

nickel /ˈnikl/ *n.* ［地］自然镍；镍　　　　　　　　　　　　　　1-A

nitrogen /ˈnaitrədʒən/ *n.* 氮　　　　　　　　　　　　　　　　1-A

nonmetallic /ˌnɒnmɪˈtælɪk/ *adj.* 非金属的　*n.* 非金属物质　　　7-A

nourish /ˈnʌriʃ/ *v.* 滋养；给营养；培育；怀有　　　　　　　　　2-A

O

oblique /əˈbliːk/ *n.* 倾斜物　*adj.* 斜的；不光明正大的　*vi.* 倾斜　8-A

obliterate /əˈblɪtəreɪt/ *vt.* 除去，消灭　　　　　　　　　　　2-A

obsidian /əbˈsɪdɪən/ *n.* 黑曜石　　　　　　　　　　　　　　　5-A

oceanography /ˌəuʃəˈnɒgrəfi/ *n.* 海洋学　　　　　　　　　　　1-B

olivine /ˈɒlɪviːn/ *n.* ［矿物］橄榄石；橄榄绿　　　　　　　　　　5-A

ore /ɔ:(r)/ *n.* 矿；矿石　　　　　　　　　　　　　　　　　　　　8-A

orthoclase /ˈɔ:θəʊkleɪz/ *n.* 正长石　　　　　　　　　　　　　　　4-B

outcrop /ˈaʊtkrɒp/ *n.* 露头；露出地面的岩层 *vi.* 露出　　　　　2-A

oversaturated /ˈəuvəˈsætʃəˌreɪtɪd/ *adj.* ［化］过饱和的；硅石的　5-A

oxidation /ɒksɪˈdeɪʃn/ *n.* ［化］氧化　　　　　　　　　　　　　　3-A

oxide /ˈɒksaɪd/ *n.* 氧化物　　　　　　　　　　　　　　　　　　5-B

P

paleontology /ˌpælɪɒnˈtɒlədʒɪ/ *n.* 古生物学　　　　　　　　　　1-B

paleoseismologist /ˌpeɪlɪəusaɪzˈmɒlədʒɪst/ *n.* 古地震学家　　　8-B

Pangaea /pænˈdʒi:ə/ *n.* 泛古陆；泛大陆　　　　　　　　　　　　2-B

paraffin /ˈpærəfɪn/ *n.* 石蜡　　　　　　　　　　　　　　　　　　7-B

parched /pɑ:tʃt/ *adj.* 焦的；炎热的；炒过的；干透的

　　　　　　　vt. 烘干；使极渴（parch 的过去分词）　　　　　3-A

pebble /ˈpebl/ *n.* 砾石，鹅卵石 *v.*（用卵石等）铺　　　　　　　3-A

percolate /ˈpɜ:kəleɪt/ *vt.* 使渗出；使过滤 *vi.* 过滤；渗出；渗透　7-A

percolating /ˈpɜ:kəleɪtɪŋ/ *n.* 渗透 *adj.* 渗透的 *vt.* 渗透（percolate 的 ing 形式）　7-A

perpendicular /ˌpɜ:pənˈdɪkjʊlə(r)/ *adj.* 垂直的，正交的；直立的　8-A

petrology /pəˈtrɒlədʒɪ/ *n.* 岩石学　　　　　　　　　　　　　　　2-A

phaneritic /ˌfænəˈrɪtɪk/ *adj.* 粗晶的；显晶岩的　　　　　　　　5-A

phosphorus /ˈfɒsfərəs/ *n.* 磷　　　　　　　　　　　　　　　　　7-B

planar /ˈpleɪnə/ *adj.* 平面的；平坦的；二维的　　　　　　　　　4-A

plutonic /plu:ˈtɒnɪk/ *adj.* 火成岩的；深成岩的　　　　　　　　　5-A

pockmark /ˈpɒkmɑ:k/ *n.* 麻子；凹坑 *vt.* 使留下痘疤；使有凹坑　3-A

potassium /pəˈtæsɪəm/ *n.* 钾　　　　　　　　　　　　　　　　　4-B

pound /paund/ *n.* 磅；英镑 *vt.* 捣碎；敲打；连续砰砰地猛击

　　　　　　　vi. 咚咚地走；（心脏）怦怦地跳　　　　　　　　2-A

precipitate /prɪˈsɪpɪteɪt/ *n.* ［化］沉淀物；*vt.*使沉淀；加速，促成；

　　　　　　　vi. ［化］沉淀；冷凝成为雨或雪等　　　　　　　　6-A

precipitation /prɪˌsɪpɪˈteɪʃn/ *n.* ［化］沉淀物；冰雹；坠落　　　6-A

prehistoric /ˌpri:hɪˈstɒrɪk/ *adj.* 史前的　　　　　　　　　　　　1-B

progressively /prəˈgresɪvli/ *adj.* 渐进地，日益增加地   2-A

prospecting /prəʊˈspektɪŋ/ *n.* 探矿；勘探   7-A

protolith /ˌprəʊtəʊˈlɪθ/ *n.* 原岩   5-B

pyroxene /paɪəˈrɒksiːn/ *n.* 辉石；锂辉石   4-B

Q

quartzite /ˈkwɒːtsaɪt/ *n.* 石英岩；硅岩   5-B

quench /kwentʃ/ *vt.* 熄灭，［机］淬火；冷却 *vi.* 熄灭；平息   5-A

R

radioactivity /ˌreɪdiəʊækˈtɪvətɪ/ *n.* 放射性；辐射   2-A

radiometric /ˌreɪdɪəʊˈmetrɪk/ *adj.* 辐射度测量的，放射性测量的   2-A

rapids /ˈræpɪdz/ *n.* ［水文］急流；湍流   3-B

reckon /ˈrekən/ *v.* 计算；认为；估计   2-A

reclamation /ˌrekləˈmeɪʃən/ *n.* （废料）回收；再利用；开垦   7-A

recur /rɪˈkɜː(r)/ *vi.* 再发生，复发   2-A

reduce /rɪˈdjuːs/ *vt.* 缩减；简化；还原   3-A

reef /riːf/ *n.* 暗礁，礁石；［地质］矿脉；收帆 *vt.* 收帆；缩帆 *vi.* 缩帆；收帆   6-A

refined /rɪˈfaɪnd/ *adj.* 精炼的；精确的；有教养的   7-B

refinery /rɪˈfaɪnərɪ/ *n.* 精炼厂；炼油厂   7-B

remains /rɪˈmeɪnz/ *n.* 遗迹；遗体   2-A

remnant /ˈremnənt/ *n.* 剩余部分 *adj.* 剩余的   3-A

rhyolite /ˈraɪəlaɪt/ *n.* ［岩］流纹岩；表面光滑的火山岩   5-A

riffle /ˈrɪfl/ *n.* 浅滩   3-B

robustness /rəʊˈbʌstnɪs/ *n.* 稳健性；健壮性   7-A

ruby /ˈruːbɪ/ *n.* 红宝石   4-A

rung /rʌŋ/ *n.* 梯级，阶梯；地位 *v.* 打电话（ring 的过去式和过去分词）   2-A

S

saltation /sælˈteɪʃ(ə)n/ *n.* 突变；（水中砂粒）跃移；不连续变异   3-B

Saudi Arabia /ˈsaudɪ əˈreɪbɪə/ *n.* 沙特阿拉伯   7-B

schist /ʃɪst/ *n.* 片岩   5-B scratch /skrætʃ/ *n.* 擦，划痕 *vt.&vi.* 抓，刮，搔　　1-A

scratchability /skrætʃəˈbɪlɪtɪ/ *n.* 柔软度　　4-A

secrete /sɪˈkriːt/ *vt.* 隐藏，藏匿；私吞；[生] 分泌　　6-A

secretion /sɪˈkriːʃn/ *n.* 分泌；分泌物；分凝　　6-A

sediment /ˈsedɪmənt/ *n.* 沉积；沉淀物　　2-A

sedimentary /ˌsedɪˈmentri/ *adj.* 沉积的；沉淀的　　5-A

seismic /ˈsaɪzmɪk/ *adj.* 地震的　　8-B

seismograph /ˈsaɪzmə(ʊ)grɑːf/ *n.* 地震仪　　1-B

shale /ʃeɪl/ *n.* [岩] 页岩　　2-A

signify /ˈsɪgnɪfaɪ/ *vt.* 表示；意味（着）　　2-A

silt /sɪlt/ *n.* 淤泥，粉砂 *vt.* 使淤塞；充塞 *vi.* 淤塞，充塞；为淤泥堵塞　　6-A

sinkhole /ˈsɪŋkhəʊl/ *n.* 落水洞；灰岩坑　　3-A

sinuous /ˈsɪnjuəs/ *adj.* 蜿蜒的；弯曲的；迂回的　　8-B

slate /sleɪt/ *n.* 板岩，高灰煤　　5-B

slickenside /ˈslɪk(ə)nsaɪd/ *n.* 岩石光滑面；镜岩　　7-B

smelting /ˈsmeltɪŋ/ *n.* [冶] 熔炼；冶炼 *v.* 精炼（smelt 的 ing 形式）　　7-A

sodium /ˈsəʊdɪəm/ *n.* [化] 钠（11 号化学元素，符号为 Na）　　4-A

solidification /səˌlɪdɪfɪˈkeɪʃn/ *n.* 凝固，固化；浓缩　　5-A

solidify /səˈlɪdɪfaɪ/ *vt.* 固化；团结；凝固 *vi.* 固化；凝固　　5-A

span /spæn/ *n.* 时距；跨度；间距；变化范围 *vt.* 延续；横跨；贯穿；遍及　　2-A

speculation /ˌspekjʊˈleɪʃn/ *n.* 推测；投机；沉思　　2-A

sphere /sfɪə(r)/ *n.* 范围，领域；球，球体　　1-A

spherical /ˈsferɪkəl/ *adj.* 球的；球面的；球状的　　2-A

spheroidal /sfɪəˈrɔɪdəl/ *adj.* 类似球体的，球状的　　6-A

stalactite /ˈstæləktaɪt/ *n.* 钟乳石　　6-B

strata /ˈstrɑːtə/ *n.*（stratum 的复数）层；地层；阶层　　6-A

stratigraphy /strəˈtɪgrəfɪ/ *n.* 地层学　　2-A

subside /səbˈsaɪd/ *vi.* 减弱，（热度）消退；沉淀　　2-A

subsystem /ˈsʌbsɪstəm/ *n.* 子系统，次系统；亚晶系　　1-A

succeeding /səkˈsiːdɪŋ/ *adj.* 随后的，以后的　　2-A

sulphide /ˈsʌlfaɪd/ *n.* 硫化物　　7-A

sulphur /ˈsʌlfə(r)/ *n.* 硫黄；硫黄色 *vt.* 使硫化；用硫黄处理	4-A
superfluous /suːˈpɜːfluəs/ *adj.* 多余的；不必要的	4-A
suspended load 悬移质	3-B
swamp /swɒmp/ *n.* 沼泽；湿地 *vt.* 使陷于沼泽；使沉没；使陷入困境	
vi. 下沉；陷入沼泽；陷入困境	6-A
symmetrical /sɪˈmetrɪkəl/ *adj.* 匀称的，对称的	8-A
syncline /ˈsɪŋklaɪn/ *n.* 向斜；向斜层	8-A
synform /ˈsɪnfɔːm/ *n.* 向形，向斜式构造	8-A
synthetic /sɪnˈθetɪk/ *adj.* 综合的；合成的；人造的 *n.* 人工合成物	4-A

T

talc /tælk/ *n.* 滑石；云母 *vt.* 用滑石粉处理；在……上撒滑石粉	4-A
tanker /ˈtæŋkə(r)/ *n.* 油轮；罐车	7-B
tar /tɑː(r)/ *n.* 焦油	7-B
tatter /ˈtætə(r)/ *vt.* 扯碎，撕碎	2-A
till /tɪl/ *n.* 冰碛	6-B
towering /ˈtaʊərɪŋ/ *adj.* 高耸的；卓越的；激烈的	3-A
traction /ˈtrækʃn/ *n.* 拖拉，牵引；拉曳	3-B
trough /trɒf/ *n.* 槽；波谷	8-A
tsunami /tsuːˈnɑːmɪ/ *n.* 海啸	8-B
twig /twɪg/ *n.* 小枝，末梢；探矿杖	6-B

U

unconsolidated /ˈʌnkənˈsɒlɪdeɪtɪd/ *adj.* 疏松的；松散的；未固结的	6-A
undermine /ˌʌndəˈmaɪn/ *vt.* 渐渐破坏	3-B
undersaturated /ˌʌndəˈsætʃəreɪtɪd/ *adj.* 未饱和的；不饱和的	5-A

V

vein /veɪn/ *n.* ［地质］岩脉；纹理；翅脉；性情	7-A
vergence /ˈvɜːdʒəns/ *n.* 趋异，构造转向；朝向	8-A
viscosity /vɪˈskɒsɪtɪ/ *n.* ［物］黏性，［物］黏度	6-A

W

wedge /wedʒ/ *n.* 楔形物　　　　　　　　　　　　　　　　　　3-A

weld /weld/ *n.* 焊接　*vt.* 焊接；使结合；使成整体点　*vi.* 焊牢　　5-A

winnow /ˈwɪnəʊ/ *n.* 扬谷；扬谷器　*vt.* 簸；把……挑出来；精选

　　　　　　　vi. 分出好坏；扬谷　　　　　　　　　　　　2-A

Appendix B

Keys to Exercises

Unit 1

Text A

Answer the following questions according to the passage you have read.

(1) The general shape of the earth is that of a flattened sphere. It is not perfectly round but slightly flattened at the poles.

(2) The earth is usually divided by scientists into three spheres: the atmosphere, the hydrosphere and the lithosphere.

(3) The solid part of the earth is called the lithosphere. We know more about the water and gases covering the earth than we know about the solid earth itself. The deepest wells extend only about 7 miles below the surface; thus, compared to the almost 8,000-mile diameter of the earth, this is only a scratch on its surface.

(4) The earth is believed to be made up of several layers; however, it can be roughly divided into crust, mantle and core (outer core and inner core) from the outer to the inner parts.

(5) No, it isn't. The mantle is thought to be composed of heavier rock than the material making up the crust.

(6) Because of the great pressure and heat in this layer, the mantle is neither quite a solid nor quite a liquid. In other words, it is thought to be in a plastic state.

(7) It is believed to be composed mostly of iron and nickel.

(8) The hydrosphere is the water layer covering the earth.

(9) The water vapor in the earth's atmosphere is one of the most essential agents responsible for many geological changes.

(10) The changes have resulted from interactions between the various interrelated internal and external earth subsystems and cycles.

1. Translate the following sentences into Chinese.

(1) 因此，相对于直径几乎达 8 000 英里的地球而言，这只不过是它的一点皮毛而已。

(2) 换言之，人们认为构成地幔的物质处于塑性状态。

(3) 科学家们估计，如果夷平所有的高山，抬升全部的海底，地球上的水也足以将地球覆盖 1 英里半深。

(4) 大气圈由气体组成，主要是围绕地球固体和液体部分的氮气与氧气。

(5) 随着时间的推移，地球的大气、海洋，以及在某种程度上其地壳的改变都受到过生命过程的影响。

2. Translate the following passage into English.

The earth is a nearly spherical planet. It has a circumference of approximately 25,000 miles (40,000 km), a polar diameter of about 7,900 miles (12,714 km), and an equatorial diameter of 7,927 miles (12,756 km).   The three main units of the earth's interior are core, mantle and crust. The diameter of the core is about 4,300 miles (6,900 km), and iron is probably its chief ingredient. The core consists of an inner part that seems solid and an outer part that appears fluid. The mantle is nearly 1,800 miles (2,900 km) thick and makes up about 84% of the volume of the earth. Since the volume of the core is about 16%, the crust actually makes up a very small part of the Earth as a whole.

Text B

Questions for Review

(1) A   (2) B   (3) D   (4) B

(5) Geology is the study of the composition, structure, and history of the earth.

(6) Geology depends upon the fundamental sciences of physics, chemistry, and biology.

(7) Chemical and X-ray analyses reveal the composition of rocks and minerals.

(8) Yes. To name only a few, there are meteorology, for the study of the atmosphere; physical geography, for the external features of the globe; oceanography, for the earth's waters and their uneven depths; soil science, for its vital and special topic.

Unit 2

Text A

Answer the following questions according to the passage you have read.

(1) Geologic time is a term used by earth scientists, referring to the earth's age or the vast span of time.

(2) He found fossil shells far inland in what are now parts of Egypt and Libya and inferred that the Mediterranean Sea had once extended much farther to the south.

(3) The concept of a spherical earth was beyond the imagination of most men at that time. / Most men didn't believe that the earth was spherical at that time.

(4) Rocks, like the pages in a long and complicated history, record the earth-shaping events and life of the past, certifying that the earth is billions of years old.

(5) The relative time scale is based on the sequence of layering of rocks and the evolution of life while the radiometric time scale is based on the natural radioactivity of chemical elements in some of the rocks.

(6) He first proposed the fundamental principle used to classify rocks according to their relative ages, concluding that each layer represented a specific interval of geologic time. He also proposed that wherever non-contorted layers were exposed, the bottom layer was the oldest layer exposed and each succeeding layer, up to the topmost one, was progressively younger.

(7) He discovered that certain layers contained fossils unlike those in other layers after cataloging fossil shells from the rocks he had collected in areas in southern England.

(8) From the results of studies, geologists can reconstruct the sequence of events that has shaped the earth's surface.

(9) The recurring events like mountain building and sea encroachment have led to the formation of the surface of the earth.

(10) They have used a fascinating mixture of words to name the divisions of geologic time.

1. Translate the following sentences into Chinese.

(1) 岩石中尘封了许多有关地球年龄的秘密，为此我们已探索了好几个世纪，也因此产生并发展壮大了地质学。

(2) 一些含有清晰可变的鱼类和其他水生动植物化石的岩层，最早是在海洋中形成的。

(3) 不管非扭曲岩层出露在哪里，底层岩层都是最早沉积的，因此也是沉积的最老岩层。

(4) 由于化石真实地记录了缓慢但延续不断的生命进程，科学家们便用它们来辨认世界各地年龄相同的岩石。

(5) 这些被岩石本身记录的旋回造山运动和循环海侵事件组成了地质时间单元，尽管我们还无法确定这些事件（发生）的确切时间。

2. Translate the following passage into English.

The geologic time is a unit of time used to describe the historic events of the earth and the time in which different geologic events take place. It has two implications: the time sequence of the events called a relative time and the time span from the occurrence of the events to the present referred to as the absolute time. The integration of the two measurements makes possible an overall understanding of the geologic events, the earth and the development of the earth crust, just based on which the geologic time scale comes into being.

Text B

Questions for Review

(1) C　　　(2) C　　　(3) B　　　(4) B　　　(5) C

(6) The Precambrian Time Span, the Paleozoic Era, the Mesozoic Era and the Cenozoic Era. Because each geologic time span, marked by some significant geologic events, bears comparatively distinctive features.

(7) In the Precambrian Time Span there was no life on the earth and at the end of this period the atmosphere was just beginning to accumulate oxygen and single celled organisms came into existence.

In the Paleozoic Era, life flourished on the earth first in the ocean,then on the land. Plants, invertebrates, vertebrates and many other new species appeared and thrived.

In the Mesozoic Era the climate was humid and tropical. Many plants and animals such as herbivores, dinosaurs and small　mammals came into existence. Birds also evolved. As dinosaurs were the dominant species for much of the era, this period is also known as "the age of dinosaurs".

In the Cenozoic Era most temperate parts of the earth were covered in glaciers and the climate got much cooler and drier than that of the Mesozoic Era. Small mammals that survived were able to grow and become dominant life on the earth. Human evolution began. All life on the earth evolved into their present day forms.

(8) (open)

Unit 3

Text A

Answer the following questions according to the passage you have read.

(1) Weathering refers to the group of destructive processes that change the physical and chemical character of rock on or near the earth's surface.

(2) Rocks and soils can be broken down through direct contact with atmospheric conditions such as heat, water, ice and pressure.

(3) Chemical weathering changes the materials that make up rocks and soil.

(4) There are 3 agents of erosion: water, wind and ice.

(5) Waves constantly crash against shores. They pound rocks into pebbles and reduce pebbles to sand. Water sometimes takes sand away from beaches. This moves the coastline farther inland.

(6) Wind is responsible for some sand dunes in some area of the Gobi Desert. It carries dust, sand, and volcanic ash from one place to another. Wind can sometimes blow sand into towering dunes.

(7) In frigid areas and on some mountaintops, glaciers move slowly downhill and across the land. As they move, they pick up everything in their path, from tiny grains of sand to huge boulders. The rocks carried by a glacier rub against the ground below, eroding both the ground and the rocks. Glaciers grind up rocks and scrape away the soil. Moving glaciers gouge out basins and form steep-sided mountain valleys.

(8) Weathering breaks down rocks that are either stationary or moving. Erosion is the picking up or physical removal of rock particles by an agent such as streams, wind or

glaciers. Weathering helps break down a solid rock into loose particles that are easily eroded. Most eroded rock particles are at least partially weathered, but rock can be eroded before it has weathered at all. A stream can erode weathered or unweathered rock fragments.

1. Translate the following sentences into Chinese.

(1) 风化可以让火成岩结合紧密的晶体变松并转变成新的矿物。

(2) 温度的变化引起岩石的扩张和收缩。当这个过程反复发生时，岩石就会松动。久而久之，岩石就裂成了碎片。

(3) 碳酸对溶解石灰岩特别有效。当碳酸渗透到位于地下的石灰岩，能撕出巨大的裂缝或挖空庞大的洞穴网。

(4) 海浪的拍击也侵蚀着海边悬崖。有时，它钻成的孔就形成了洞穴。

(5) 地球的岩石，在风化和侵蚀中被慢慢地雕琢、打磨和抛光，演变成前所未有的艺术品，剩余的残迹随后被冲入大海。

2. Translate the following passage into English.

Global warming, the increase in temperature around the world, is speeding erosion. The change in climate has been linked to more frequent and more severe storms. Storm surges following/in the wake of hurricanes and typhoons threaten to erode miles of coastline and coastal habitat. Homes, businesses, and economically important industries, such as fisheries are settled down along these costal areas. The rise in temperature is also quickly melting glaciers, which causes the sea level to rise faster than organisms can adapt to it. The rising sea erodes beaches much more quickly. It is estimated that a rise in sea level of 8 to 10 centimeters will lead to enough erosion to threaten the safety of buildings, sewer systems, roads, and tunnels.

Text B

Questions for Review

(1) D　　　(2) B　　　(3) C　　　(4) A　　　(5) B

(6) This is due to the river not having a lot of spare energy as it is using 90% of its energy to overcome obstacles such as large rocks and boulders.

(7) They are rapids, small meanders, small floodplains, pools and riffles.

(8) The type of transport taking place depends on the size of the sediment and the amount of energy that is available to undertake the transport.

Unit 4

Text A

Answer the following questions according to the passage you have read.

(1) A mineral is a naturally occurring, inorganic, crystalline solid, with a narrowly defined chemical composition and characteristic physical properties.

(2) No, because they are naturally occurring substances not laboratory products.

(3) The physical properties of minerals are determined by their chemical composition and by the geometric arrangement of the atoms composing them.

(4) No, because some organisms, including corals, clams, and a number of other animals, construct their shells of the compound calcium carbonate (C_aCO_3), which is either the mineral aragonite or calcite, but we can't call them minerals.

(5) No, not all rigid substances are crystalline solids and crystalline structure can be demonstrated even in minerals lacking obvious crystals.

(6) Mineral composition is generally shown by a chemical formula.

(7) The physical properties of a mineral are controlled by composition and structure.

(8) No, because color is also apt to be the most ambiguous of physical properties. Color is extremely variable in quartz and many other minerals because even minute chemical impurities can strongly influence it.

(9) Most minerals are able to develop their characteristic crystal faces only if they are surrounded by a fluid that can be easily displaced as the crystal grows.

(10) A mineral may be defined as a naturally occurring, inorganic and crystalline solid with a fairly definite chemical composition and characteristic physical properties.

1. Translate the following sentences into Chinese.

(1) 因为它们是天然生成物，不是实验室里的产品，所以只有在极少数情况下它们才是纯净的化合物。

(2) 矿物还具有一定的物理性质，这取决于它们的化学成分和组成矿物的原子的几何排列。

(3) 换言之，一个结构良好矿物晶体的规则几何形状是其内部有序的原子排列的外在表现。

(4) 颜色在石英和很多其他矿物中是极大的一个变量，因为即使是很细微的化学杂质也会对其产生极强的影响。

(5) 大多数岩石沉积都含有金属或矿物，但如有用矿物或金属的浓度低得难以证明其有开采价值，它们则被认为是废石或脉石矿物。

2. Translate the following passage into English.

Since one of the critical differences between minerals and rocks is that minerals are approximately homogeneous substances, and most rocks are not, this means that one piece of quartz will be about as hard as another piece, that it will have the same specific gravity, and if formed in a similar environment, it will have about the same crystal form.

Text B

Questions for Review

(1) D　　　　(2) B　　　　(3) B　　　　(4) B

(5) For one thing, many combinations of elements are chemically impossible; no compounds are composed of only potassium and sodium or of silicon and iron, for example. Another important factor restricting the number of common minerals is that only eight chemical elements make up the bulk of the earth's crust.

(6) Minerals are grouped with similar crystal structures and compositions.

(7) 6 groups are mentioned here including feldspar group, pyroxene group, amphibole group, mica group, calcite and quartz respectively.

Unit 5

Text A

Answer the following questions according to the passage you have read.

(1) Igneous rocks are produced by cooling and solidification at molten rock-making

material called magma or lava.

(2) Magma is molten rock that is formed in very hot conditions inside the earth with or without suspended crystal and gas bubbles; while lava is the very hot liquid rock that comes out of a volcano to reach the surface.

(3) Intrusive rocks are formed from magma that cools and solidifies within the crust of a planet. They are coarse-grained, younger than the rocks they intrude and exposed in places at the surface today. Extrusive igneous rocks, formed at the crust's surface as a result of the partial melting of rocks within the mantle and crust, are mostly finely grained.

(4) Intrusive rocks are formed from magma that cools and solidifies within the crust of a planet; while extrusive igneous rocks are formed at the crust's surface as a result of the partial melting of rocks within the mantle and crust. Intrusive rocks are coarse-grained because magma cools slowly while extrusive rocks are finely grained because of quick cooling and solidification.

(5) Yes, because granite is the chemical and mineralogical equivalent of rhyolite.

(6) It is called plutonic rock.

(7) Igneous rocks can additionally be classified according to mode of occurrence, texture, mineralogy, chemical composition and the geometry of the igneous body.

(8) Because feldspathoids cannot coexist in a stable association with quartz.

(9) Igneous rocks which have crystals large enough to be seen by the naked eye are called phaneritic.

(10) Generally speaking, phaneritic implies an intrusive origin while aphanitic an extrusive one.

1. Translate the following sentences into Chinese.

(1) 岩浆可以源自地幔或地壳先存岩石的局部熔体。

(2) 溶解气是岩浆和熔岩的重要组分，但在凝固过程中倾向于从造岩矿物中排出。

(3) 花岗岩是一种主要由长石和石英构成的粗粒岩石，其在化学成分和矿物成分上和流纹岩相同，同时也是在陆地上发现得最多的侵入岩。

(4) 喷出岩包括喷出地表或近地表处由热熔岩冷却形成的细粒或玻璃质岩石，以及火山爆发过程中喷射到空中的火山灰和玻璃焊接（熔结）的碎片构成的岩石。

(5) 在简单的分类方案中，可以根据岩石中长石的类型、是否存在石英和岩石中没有长石或石英及岩石中铁、镁矿物的类型将火成岩（岩浆岩）进行分类。

2. Translate the following passage into English.

Igneous rocks, also called magmatic rock and derived from the Latin word *ignis* meaning fire, are formed underground or on the surface of the earth through the cooling and solidification of molten magma or lava at depth. Igneous rocks can be grouped into intrusive and extrusive rocks. Over 700 types of igneous rocks have been found, most of them having formed beneath the surface of the earth's crust. Granite, andesite and basalt are familiar examples. Generally, igneous rocks are likely to occur in volcanic field along the margin of some continental plates.

Text B

Questions for Review

(1) B

(2) Heat, pressure and chemical reaction of solutions.

(3) Metamorphism is characterized by growth of new minerals from pre-existing minerals through recrystallization and deformation of existing minerals either in shape or orientation.

(4) The metamorphic rock is distinct both mineralogically and texturally from the parent rock.

(5) They may be formed simply by being deep beneath the earth's surface, subjected to high temperatures and great pressure of the rock layers above it. They can form from tectonic processes such as continental collisions, which cause horizontal pressure, friction and distortion. They are also formed when rock is heated up by the intrusion of hot molten rock called magma from the earth's interior.

(6) A metamorphic rock owes its characteristic texture and particular mineral content to several factors, the most important being (1) the composition of the parent rock before metamorphism; (2) temperature and pressure during metamorphism; (3) the effects of tectonic forces; and (4) the effects of fluids, such as water.

(7) Such factors as pressure and the presence or absence of other substances will affect the stability temperature range of a mineral.

Unit 6

Text A

Answer the following questions according to the passage you have read.

(1) Sediment is the collective name for loose, solid particles, which means it's unconsolidated and the grains are separate or unattached to one another.

(2) Those loose, solid particles originate from: (1) weathering and erosion of preexisting rocks; (2) chemical precipitation from solution, including secretion by organism in water.

(3) Because it yields particles and dissolves substances, both of which might be raw materials for sedimentary rocks.

(4) Solid particles derived by mechanical and chemical weathering, minerals precipitated from solution by chemical processes, or minerals secreted by organisms when they build their skeletons.

(5) Weathering, erosion and sediment transport.

(6) There are four ways of sediment transport are mentioned here which are glaciers, wind, waves and marine, and running water respectively.

(7) Because of its high viscosity and manner of flow, a glacier does not sort the sediment it carries and deposits all sediment sizes in the same place.

(8) Deposition occurs when running water, glacial ice, waves, or wind loses energy and can no longer transport its load; besides, it also refers to the accumulation of chemical or organic sediment.

1. Translate the following sentences into Chinese.

(1) 所有的沉积岩均由沉积物构成。沉积物是由机械风化和化学风化形成的固态颗粒，由化学过程从溶液中沉淀形成的矿物或当生物形成骨架时有机质分泌形成的矿物。

(2) 其一，它只是简单的尺寸标示；其二，它也可表示被称为黏土矿物的某些片状硅酸盐。然而，大部分黏土矿物和黏土的粒度一样。

(3) 磨圆就是在运移过程中磨掉岩屑的尖利棱角。

(4) 黏土和泥沙的小颗粒可能会被搬运到没有流动水的湖里，在那里沉淀下来并形成一层淤泥。同样，淤泥可能会沉积在一条溪流的漫滩上。

(5) 化石为它们所在的岩层"贴上了标签"，并提供许多关于过去的信息。

2. Translate the following passage into English.

The sediments have been transported and dropped or precipitated by such geologic agents as running water, ocean waves, ocean currents, wind and ice. Some of the fragments may be rounded; others are angular. Such features depend upon the distance of transportation and other factors.

Text B

Questions for Review

(1) C　　(2) B　　(3) B

(4) Clastic sedimentary rocks are formed from cemented sediment grains that are fragments of preexisting rocks.

(5) It can be distinguished from breccia by the definite roundness of its particles.

(6) Because grains are rounded so rapidly during transport, it is unlikely that the angular fragments within breccia have moved very far from their source.

(7) Sandstone is formed by the cementation of sand grains and any deposit of sand can lithify to sandstone.

Unit 7

Text A

Answer the following questions according to the passage you have read.

(1) Mineral deposits, denoting a concentration of useful minerals, include both ores and nonmetallic minerals.

(2) Most ore deposits are named according to either their location, or after a discoverer, or after a historical figure, a prominent person, something from mythology or the code name of the resource company which found it.

(3) The common types of ores are gold, silver, copper, lead and iron.

(4) Yes. Silver and copper may occur native, but in combination with other elements as

well. For example, copper occurs along with oxygen as an oxide, or along with sulphur as a sulphide.

(5) The first class is primary ores that are found in the positions in which they were originally formed, and the other secondary ores have been transported from their original position by some agency.

(6) The secondary ores came into the rock after the rock had been formed and so are deposited in some kind of cavity or crack.

(7) The fissure fillings, called fissure veins as well are the chief source of gold, silver and copper.

(8) Because those easily accessible ore deposits close to the earth's surface have already been exploited by humans in the past.

(9) Conducting a pre-feasibility study is to determine the theoretical economics of the ore deposits, which indicates whether further investment in estimation and engineering studies are warranted and identifies key risks and areas for further work.

(10) Geologists should draw up a mine plan to evaluate the economically recoverable portion of the deposit, the metallurgy and ore recoverability, marketability and payability of the ore concentrates, engineering, milling and infrastructure costs, finance and equity requirements and a cradle to grave analysis of the possible mine, from the initial excavation all the way through to reclamation.

1. Translate the following sentences into Chinese.

(1) 术语"矿床"是指有用矿物的集合。

(2) 从技术上讲，矿石里也混有无价值的物质，被称为"脉石"，可通过开采获取一定经济效益。

(3) 第一种是原生矿石，它们发现于最初形成的地方。第二种叫次生矿，是从最初形成的位置通过一定介质运移的矿物。

(4) 我们知道，许多狭长裂缝严重地将形成地壳的岩石切割撕裂。

(5) 随着时间的推移，今后开发矿床将更加困难，因为那些接近地表，相对较容易发现的矿床这些年都已经被开发了。

2. Translate the following passage into English.

Ore deposits, formed by geological process, are the concentration of useful ores which are valuable for exploitation. They are the result of geological process; however, unlike those general rocks, they have economic value. With the development of science and technology, and with the decrease of mining processing costs, many ore deposits with low contents have been overexploited.

Text B

Questions for Review

(1) Petroleum fields were created by the remains of animal and plant life being compressed on the sea bed by billions of tons of silt and sand several million years ago.

(2) When small sea plants and animals die they will sink, then lie on the sea bed where they will decompose and mix with sand and silt. During the decomposition process tiny bacteria will clean the remains of certain chemicals such as phosphorus, nitrogen and oxygen. This leaves the remains consisting of mainly carbon and hydrogen. At the bottom of the ocean there is insufficient oxygen for the corpse to decompose entirely. The partially decomposed remains will form a large, gelatinous mass, which will then slowly become covered by multiple layers of sand, silt and mud. This burying process takes millions of years, with layers piling up one atop another. Finally, when the depth of the buried decomposing layer reaches somewhere around 10,000 feet the natural heat of the earth and the intense pressure will combine to act upon the mass. The end result, over time, is the formation of petroleum.

(3) In its strictest sense, petroleum includes only crude oil. Crude oil is the natural form in which petroleum is first collected, and clear, green or black and may be either thin like gasoline or thick like tar. But in common usage it includes all liquid, gaseous, and solid hydrocarbons.

(4) Petroleum contains carbon, hydrogen, nitrogen, oxygen and sulphur with a few trace metals making up a very small percentage of its composition.

(5) Denser petroleum composition with a less flammable level of hydrocarbons and sulphur are expensive to refine into a fuel and are therefore more suitable for plastics manufacturing.

(6) The petroleum industry generally classifies crude oil by the geographic location it is produced in, its API (American Petroleum Institute) Gravity (an oil industry measure of density), and its sulphur content.

(7) Because sweet oil has fewer environmental problems and requires less refining to meet sulphur standards imposed on fuels in consuming countries.

(8) There are several major oil producing regions around the world. The Kuwait and Saudi Arabia's oil fields are the largest, although Middle East oil from other countries in the region such as Iran and Iraq also make up a significant part of world production.

(9) Petroleum was used as a lighting fuel, once it had been distilled, and turned into kerosene.

(10) The largest share of oil products is used as "energy carriers", i.e. various grades of fuel oil and gasoline.

Unit 8

Text A

Answer the following questions according to the passage you have read.

(1) There are six terms to describe the components of folds. They are core, limb, hinge, axial plane, fold profile and inflexion point.

(2) Folds can be classified by fold shape, tightness, dip or the axial plane, and the bending of single folded layer

(3) If the core of an antiform contains the oldest rocks, it is an antiformal anticline, while if the core of an antiform contains younger rocks, it is an antiformal syncline.

(4) The angle between fold's limbs is called interlimb angle, which is used to define fold tightness. Gentle folds have an interlimb angle of between 180° and 120°; open folds range from 120° to 70°, close folds from 70° to 30° and tight folds from 30° to 0°. Isoclines, or isoclinal folds, have an interlimb angle of between 10° and 0°, with essentially parallel limbs.

(5) We use enveloping surface concept to judge the symmetry of fold: if enveloping

surface and axial surface are nearly perpendicular, the fold is said to be symmetrical; otherwise it is asymmetrical.

(6) There are four types of folds. They are upright fold, inclined fold, overturned fold and recumbent fold.

(7) (open) The study on folds is very important to show the deformation history of certain regions. Folds are also closely related to the formation of ores. At the same time, folds are a very important factor influencing the engineering geological environments.

1. Translate the following sentences into Chinese.

(1) 褶皱是岩石中最常见的地质结构之一，可以构成一些最壮观的地貌特征。

(2) 褶皱弯曲面上最大弯曲点的连线叫褶皱的枢纽。

(3) 岩石的褶皱说明它是塑性应变而非弹性或脆性应变。

(4) 另一方面，如果背形构造的核部为较新岩层，它被称为背形向斜。如果向形构造的核部为较老岩层，则其被称为向形背斜。

(5) 直立褶皱有一个几乎垂直的轴面，两翼岩层倾向相反。

2. Translate the following passage into English.

Fold mountains are formed when two of the tectonic plates that make up the earth's crust push together at their border. The extreme pressure forces the edges of the plates to buckle and upwards into a series of folds. Fold mountains are created through a process called "orogeny." An orogenic event takes millions of years to create a mountain range because tectonic plates move only centimeters every year. Fold mountains are the most common type of mountain on the earth. Other types of mountains are volcanic mountains, erosional mountains, and fault-block mountains. Volcanoes create volcanic mountains. Erosional mountains are produced as wind and water wear away soft portions of land and leave rocky hills. Fault-block mountains are formed where parts of continental crust are displaced.

Text B

Questions for Review

(1) B　　　(2) A　　　(3) C　　　(4) B　　　(5) D

(6) The slip along a fault describes the movement parallel to the fault plane. We speak of dip slip where movement is down or up parallel to the dip direction of the fault. The term strike-slip applies where movement is parallel to the strike of the fault plane.

(7) A. the normal fault

    B. the thrust fault

    C. the strike-slip fault

    D. the oblique-slip fault

(8) First, a fault's discontinuity has a large influence on the mechanical behavior of soil and rock masses in tunnel, foundation, or slope construction. Second, the level of a fault's activity can be critical for locating buildings, tanks, and pipelines and assessing the seismic shaking and tsunami hazard to infrastructure and people in the vicinity. Third, radiocarbon dating of organic material buried next to or over a fault shear is often critical in distinguishing active from inactive faults, which is helpful to estimate the rough future fault activity.

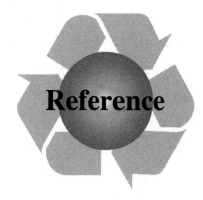

Reference

[1] CHARLES C. Plummer & David McGeary Physical Geology [M]. 7th ed. Wm C. Brown Publishers, 1996.

[2] CONDIE K C. Plate Tectonics and Crustal Evolution [M]. 4th ed. Butterworth-Heinemann, 2010.

[3] MEARS, BRAINERD JR. Essentials of Geology[M]. Litton Educational Publishing, Inc., 1978.

[4] ORESKES. Naomi Plate Tectonics: An Insider's History of the Modern Theory of the Earth[M]. Westview, 2003.

[5] SHANNON. Naylor Petroleum Basin Studies[M]. Graham & Trotman Limited, 1989.

[6] WEGENER. Alfred The Origin of Continents and Oceans[M]. Courier Dover,1966.

[7] 秦荻辉. 科技英语写作[M]. 北京：外语教学与研究出版社，2007.

[8] 王铭和. 英语科技论文写作[M]. 青岛：中国海洋大学出版社，2003.

[9] 周光文. 科技英语翻译和写作[M]. 北京：科学出版社，2015.

反侵权盗版声明

电子工业出版社依法对本作品享有专有出版权。任何未经权利人书面许可，复制、销售或通过信息网络传播本作品的行为；歪曲、篡改、剽窃本作品的行为，均违反《中华人民共和国著作权法》，其行为人应承担相应的民事责任和行政责任，构成犯罪的，将被依法追究刑事责任。

为了维护市场秩序，保护权利人的合法权益，我社将依法查处和打击侵权盗版的单位和个人。欢迎社会各界人士积极举报侵权盗版行为，本社将奖励举报有功人员，并保证举报人的信息不被泄露。

举报电话：（010）88254396；（010）88258888

传　　真：（010）88254397

E-mail：　dbqq@phei.com.cn

通信地址：北京市海淀区万寿路 173 信箱
　　　　　电子工业出版社总编办公室

邮　　编：100036